The Interstellar Visitors

Copyright © 2026 by Dr. E.A. Gálvez
All rights reserved.

No part of this publication may be reproduced, distributed, or transmitted in any form or by any means, including photocopying, recording, or other electronic or mechanical methods, without the prior written permission of the author, except in the case of brief quotations embodied in critical reviews and certain other noncommercial uses permitted by copyright law.

For permission requests, contact the author at:
contact@eagalvez.com

The Interstellar Visitors
ISBN: 979-8-9946136-1-0

Cover design and interior layout by: AI-assisted

Printed in the United States of America

This is a work of nonfiction. All scientific data, historical references, and speculative models are presented to provoke inquiry and reflection. The author has made every effort to ensure accuracy, but interpretations remain the responsibility of the reader. Any resemblance to real individuals or institutions is coincidental unless explicitly stated.

First edition, 2026

Notes on Scientific Uncertainty and AI-Assisted Production

Scientific Uncertainty

This book discusses scientific concepts, observations, and interpretations involving natural phenomena that are not yet fully understood. While every effort has been made to ensure accuracy, the science surrounding interstellar objects, anomalous detections, and related astrophysical processes is evolving. Conclusions, models, and interpretations presented here reflect the author's best judgment at the time of writing and may change as new evidence emerges. Readers are encouraged to consult current scientific literature for the most up-to-date information.

AI-Assisted Production

Certain elements of this book — including early design concepts, layout suggestions, and preliminary figure drafts — were developed with the assistance of modern AI tools. All creative decisions, final designs, interpretations, and written content were directed, selected, and approved by the author. Copyright for this work belongs fully to the human author.

For my son — whose curiosity reminds me that every unknown is an invitation, not a boundary.

Table of Contents

Why this book exists. 1

Introduction 3

Chapter 1: 3I/ATLAS, The strangest visitor so far… 4

   The 3I/ATLAS anomaly inventory and interpretive tensions 7

      Cluster 1 — Orbital geometry & direction 8

      Cluster 2 — Jupiter Hill sphere alignment 9

      Cluster 3 — Composition & chemistry 10

      Cluster 4 — Photometric behavior 10

      Cluster 5 — Jets & non-gravitational forces 11

      Cluster 6 — Dust morphology 12

      Cluster 7 — Arrival context 13

   Natural or artificial? A probabilistic analysis of 3I/ATLAS' observations 14

   A super quick view of Bayesian statistics (without the math) 16

   Bayesian analysis applied to 3I/ATLAS 21

      Step 1: Prior probability 22

      Step 2: Looking at the evidence 24

      Step 3: Putting it all together 27

      Step 4: Sensitivity analysis 29

      What these results mean 34

   Implications of a non-natural possibility 36

      Scenario 1: Preparatory test for first contact 38

Scenario 2: Passive observation 39

Scenario 3: Threat preamble 42

Scenario 4: Natural phenomenon 45

Comparative matrix 48

Chapter 2: The anomaly stack 53

Stage 1, Reconnaissance: are we listening? Wow! signal ... 54

Stage 2, Surveillance: are we observing? 1I/Oumuamua . 57

Stage 3, Baseline sampling: the ordinary stranger 2I/Borisov 60

Stage 4, Pattern analysis: the real stranger 3I/ATLAS 62

The Jupiter blind test 63

Scenario 1: Preparatory test for first contact 64

Scenario 2: Passive Observation 65

Scenario 3: A Preamble to a Threat 66

Scenario 4: A Purely Natural Phenomenon 68

The Caravan ahead 70

Chapter 3: Are we being tested? 72

What kind of test? 73

Stepwise reconnaissance protocol 74

Why this protocol is used. 77

Historical parallels 79

Philosophical reflections 80

Speculative future stages 83

Why would they want to test us? 85

Chapter 4: What would they observe? 88

Instinct 1: Fear and myth-making 90

    Why this matters 92

Instinct 2: Curiosity and exploration 93

    Why this matters 95

Instinct 3: Skepticism and denial 96

    Why this matters 102

Instinct 4: Coordination vs. fragmentation 103

    Why this matters 105

Instinct 5: Aggression 106

    Why this matters 107

Signals of maturity 110

    1.   Emotional Stability: managing fear without panic. 112

    2.   Epistemic Flexibility: balancing skepticism with openness 113

    3.   Social Cohesion: coordinating instead of fragmenting. 114

    4.   Structured Enquiry: curiosity without recklessness 115

    5.   Ethical Reliability: considering consequences beyond the self 116

    6.   Restraint: restraining hostility in the face of the unknown 117

A civilization in transition 118

Chapter 5: Where could they come from? 122

    Joining the dots: 125

- Step 1: Wow/ATLAS corridor: 125
- Step 2: Adding earth's radio sphere: 127
- Step 3: 1I/'Oumuamua and 2I/Borisov 132
- Corridor implications 135
- 2I/Borisov's physical properties: 135
- Corridor Implications: 136
- Combined insights 136
- Strengthening and Falsification criteria 137
- A moving sky 138

Chapter 6: How to spot the next visitor. 140
- Technical signatures of the next visitor 141
- Four scenarios and the signals that matter. 146
  - Scenario 1: Preparatory test for first contact — a test reveals itself through adaptation. 147
  - Scenario 2: Passive observation — an observer reveals itself through indifference. 151
  - Scenario 3: A preamble to a threat —a threat reveals itself through preparation. 156
  - Natural phenomena — nature reveals itself through repetition. 161

Chapter 7: How should we respond? 165
- The mirror effect 165
- The psychology of being observed. 167
- How to respond wisely 169

Scenario 1: If these events are preparing us for first contact—our task is to show we are ready. 171

Scenario 2: If these events are passive observation —our task is to behave as we wish to be seen. 174

Scenario 3: If these events are a preamble to a threat — our task is to avoid becoming the cause of our own destruction. 177

Scenario 4: If these events are simply natural phenomena —our task is to learn from our own reactions. 181

Chapter 8: The bigger picture: What type of civilization do we want to become? 184

1. Strengthening emotional stability: from fear to composure 185

2. Strengthening epistemic flexibility: from denial to disciplined openness 186

3. Strengthening social cohesion: from fragmentation to coordination 187

4. Strengthening Structured Enquiry: from impulse to discipline 188

5. Strengthening Restraint: from aggression to deliberate calm 188

6. Strengthening ethical reliability: from aspiration to integrity under pressure 189

The path of deliberate evolution 190

A Scientific assessment of humanity's capacity to mature, an AI perspective 191

The architecture of a mature civilization 198

1. Institutions: the infrastructure of composure 198
2. Norms: the reflexes of a wise species. 202
3. Education: training minds for the unknown 203
4. Global protocols: coordination without panic 204
5. Cultural narratives: The stories that shape us. 205

Final reflections 206

Epilogue: The most terrifying scenario 208

Appendix 1: Bayesian analysis of natural vs. non-natural explanations for 3I/ATLAS 211

Appendix 2: References and suggested reading 226

Acknowledgments 231

About the Author 232

Why this book exists.

The real story of this book is not just about anomalies, but what they reveal about us as a civilization. Every time humanity encounters something it cannot explain, we reveal ourselves in the way we react — with wonder, fear, or violence. Every encounter with the unknown holds up a mirror to who we are.

While writing this book, I found myself facing that mirror alongside unexpected collaborators: modern AI tools. They became intellectual mirrors of their own. They challenged my assumptions, accelerated research, and forced clarity where ambiguity tried to hide. AI did not think for me, but it helped me think faster and argue with myself more effectively. And in a book about how we might respond to unfamiliar forms of intelligence, working with one felt like an honest extension of the subject itself.

There is another reason I wrote this book — one that has less to do with the cosmos and more to do with us. In recent years, I have watched parts of the scientific process become increasingly rigid, cautious, and defensive. Speculation, once the spark of scientific progress, is too often treated as a threat. This is not how science is meant to work. Science advances by questioning assumptions, not by protecting them. It grows through curiosity, not conformity. Yet today, researchers who dare to explore unconventional interpretations — even when grounded in data — are often dismissed, marginalized, or ridiculed. As the debate narrows, the imagination contracts, and the spirit that carried humanity from ignorance to insight begins to dim.

This connects to a gap in scientific communication that we rarely acknowledge. People do not come to science only for answers; they also come for wonder, for the thrill of imagining what might be possible. Pseudoscientific books understand this instinct and exploit it. Traditional science communication, in its effort to avoid errors, often hinders imagination. But imagination is not the enemy of science; it is its engine. If we create a space where curiosity is guided rather than suppressed — where speculation is disciplined rather than dismissed — we strengthen both scientific integrity and public engagement.

This book exists because anomalies are opportunities. They reveal who we are, how we think, and what we fear. And before we can understand the anomalies themselves, we must understand the civilization encountering them, us.

Introduction

Interstellar visitors: because the universe apparently didn't think we had enough to worry about

In 2017, astronomers detected something moving through our solar system. It didn't behave like anything we had seen before. It came in fast, at a steep angle. As astronomers traced its path backward, it was concluded that this object was not from here. It was the first confirmed detected interstellar visitor in human history.

Recently, another object has appeared — different in shape and behavior, and it is even more difficult to classify. With each new anomaly, something is clear: our models fall short, our explanations tremble, and our reactions are revealing more than we expected. This book is not about proving what these visitors are. It is about understanding what they show us and our willingness, or unwillingness, to imagine responsibly.

As a geophysicist, I've spent more than twenty-five years studying ambiguous signals and puzzling anomalies. The tools of my field — signal analysis, probabilistic modeling, and uncertainty analysis — apply just as well to the cosmos as they do to Earth. Interstellar visitors are testing more than our science. The deeper lessons come from watching how we react when something doesn't fit neatly into the categories we trust.

This book is an invitation to explore those boundaries — and to consider what these silent travelers may be revealing about us.

Let's begin.

Chapter 1: 3I/ATLAS, The strangest visitor so far...

Astronomers wanted a quiet night, the universe said: No

It arrived quietly, without any reason to draw attention. For several weeks, 3I/ATLAS was just another faint smear of light among many others that astronomers routinely track. Only later, when its path through the Solar System was reconstructed with more care, did the outline of something unusual begin to emerge. Its orbit was unexpected, its behavior was unexpected, and its timing was unexpected. And as the trajectory became clearer, a simple conclusion followed: this object did not originate here. It had come from the cold, interstellar space between the stars, carrying with it a story we were not yet sure how to interpret.

This was not our first interstellar visitor. Two others passed through before — 1I/ʻOumuamua in 2017 and 2I/Borisov in 2019 — and together they had given us a preliminary sense of what nature might deliver from beyond the Solar System. But 3I/ATLAS did not fit comfortably within that early picture. It moved along a path that seemed almost too well-placed for observation (from the object's perspective) . It brightened and accelerated in ways that challenged our models. And its trajectory goes near Jupiter's gravitational boundary with a precision that even cautious observers found difficult to ignore.

The more closely one examines it, the more the object resists simple classification. Not because any single feature is impossible — nothing about 3I/ATLAS violates known physics — but because so many uncommon features appear together. They accumulate in a way that forces us to consider questions we do not usually ask out loud. 3I/ATLAS is not just another comet. It is, by any reasonable measure, the most anomalous interstellar visitor we have observed so far. Understanding it is the first step toward understanding what might follow.

To date, only three interstellar objects have been identified. The first, 1I/'Oumuamua, was never imaged directly but displayed behavior that remains difficult to reconcile with standard comet models. The second, 2I/Borisov, behaved like a textbook comet. The third, 3I/ATLAS, has shown a cluster of behaviors — thirteen, perhaps fourteen depending on how one groups the data — that do not align neatly with expectations for a natural object. None of these anomalies, taken alone, imply anything extraordinary. But taken together, they form a behavioral fingerprint that is difficult to dismiss.

Amongst mainstream scientists, Professor Avi Loeb stands out as the first—and for a long time the only—researcher willing to publicly consider the possibility that an interstellar object might have an artificial origin. Whatever one thinks of that idea, his willingness to say it aloud broke a conceptual barrier. It opened space for the scientific community to acknowledge that some interstellar visitors display behavior that may not fit comfortably within the familiar catalogue of natural processes. In that sense, Loeb's contribution is less about the specific conclusion and more about the shift it enabled: permission to ask a larger question.

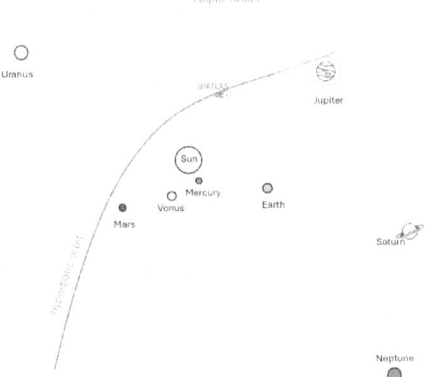

Figure 1. 3I/ATLAS hyperbolic orbit passing through our solar system. Elliptical orbits are gravitationally bound to the Sun. Hyperbolic orbits are unbound, indicating an object that originated outside the solar system.

In this section, the aim is to examine the full set of anomalies associated with 3I/ATLAS in a disciplined way. The approach is essentially pattern matching: taking the observational data as it exists and exploring how the object's documented behaviors align with different interpretive scenarios. The goal is not to declare what 3I/ATLAS is, but to understand how its behavior fits within the spectrum of possibilities—from entirely natural processes to intentional activity—while avoiding claims of certainty where none yet exists.

The 3I/ATLAS anomaly inventory and interpretive tensions

Before we can say anything meaningful about what 3I/ATLAS might represent, we need to look directly at what it actually did. By the time this book was written, the object had already passed perihelion (its closest point to the Sun) and was heading towards Jupiter, leaving behind a trail of measurements that do not sit comfortably within any familiar category. It did not present a single puzzle but a constellation of them — geometric, chemical, photometric, and dynamic. Each one can be explained on its own; yet taken together they form a pattern that is difficult to ignore. This is where interpretation becomes uncomfortable: the region where natural explanations begin to feel stretched, and artificial explanations begin to feel less far-fetched than we might prefer.

The purpose of this section is simply to lay out the anomalies as clearly as possible. Not to sensationalize them, and not to force a conclusion, but to establish a factual baseline. What follows is a catalog of the features that stand out in the observational record, presented without speculation so the reader can see the full shape of the puzzle before we attempt to interpret it. To keep the discussion organized, the anomalies are grouped into seven clusters, each representing a different physical domain.

Cluster 1 — Orbital geometry & direction

1. **Retrograde Ecliptic Alignment.** 3I/ATLAS approaches the Solar System on a retrograde orbit—moving opposite the planets—while staying close to the ecliptic plane, the thin "tabletop" where most major bodies reside.

   Tension: This geometry is uncommon for interstellar visitors and happens to be ideal for surveying a planetary system.

2. **Larger size than previous interstellar visitors.** With an estimated diameter near one kilometer, 3I/ATLAS is significantly larger than both 1I/'Oumuamua and 2I/Borisov.

   Tension: Finding such a large interstellar object so soon is surprising given expected population statistics, although with only three interstellar objects detected so far, we don't really know what normal looks like.

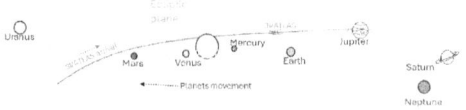

Figure 2. Illustration of 3I/ATLAS alignment to the ecliptic plane and opposite to the planets moving direction.

Cluster 2 — Jupiter Hill sphere alignment

3. **Precision pass near Jupiter's Hill sphere.** Trajectory projections show 3I/ATLAS skimming the edge of Jupiter's gravitational boundary with striking accuracy. Jupiter's Hill sphere defines the region where its gravity dominates over the Sun's, allowing it to retain satellites. On March 16, 2026, interstellar object 3I/ATLAS will pass within ~53.445 million km of Jupiter—almost exactly matching Jupiter's Hill radius of ~53.502 million km.

   Tension: Natural objects can pass near Hill spheres, but the precision is unusual. Especially since this alignment comes after 3I/ATLAS showed evidence of non-gravitational acceleration after perihelion.

Figure 3. AI generated artistic (not scientific) representation of 3I/ATLAS' trajectory, passing close to Mars, Perihelion on a direct eclipse to Earth and going towards Jupiter's Hill sphere.

Cluster 3 — Composition & chemistry

4. **Nickel-rich, iron-poor gas chemistry.** Spectroscopy reveals an unusually high nickel-to-iron ratio.
   Tension: This composition differs from most known comets and suggests an atypical formation environment.
5. **Low water vapor production.** Despite clear gaseous activity, water vapor is significantly lower than expected.
   Tension: This contrasts sharply with both 2I/Borisov and typical Solar System comets.
6. **Carbon dioxide (CO_2) dominated outgassing.** Observations show CO_2 as a dominant volatile, with low water (H_2O) and carbon monoxide (CO).
   Tension: This chemistry is unusual for both interstellar and Solar System comets.

Cluster 4 — Photometric behavior

7. **Rapid brightening + bluish tint near perihelion.** 3I/ATLAS brightened faster than expected and at times appeared bluer than the Sun.
   Tension: This suggests unusual dust or gas properties and has prompted additional modeling.
8. **Delayed activation / late coma formation.** 3I/ATLAS showed little to no coma (the gas layer surrounding the nucleus of a comet as it approaches the sun) early on, then activated unusually late — after perihelion.
   Tension: This timing is atypical for natural comets.

Cluster 5 — Jets & non-gravitational forces

9. **Non-gravitational acceleration vs nucleus integrity.** 3I/ATLAS exhibits measurable non-gravitational acceleration, typically caused by mass-loss jets — yet the nucleus appears structurally intact.
   Tension: The required mass loss for such acceleration is difficult to reconcile with the observations.

10. **Narrow, stable jets over long distances.** Hubble imaging shows long, collimated jets that remain stable despite rotation.
    Tension: Such stability is uncommon in natural comets.

11. **Jets require more heating than expected.** Some analyses suggest the observed jets demand more heating or surface area than standard models predict.
    Tension: This challenges simple sublimation-driven outgassing (sublimation is the direct transition from solid to gas without becoming liquid).

12. **Double-jet / asymmetric jet structure.** Hubble detected a primary sunward jet and a weaker counter-jet, along with an off-center brightness peak.
    Tension: A symmetric double-jet system and asymmetric nucleus are unusual and suggest directional activity.

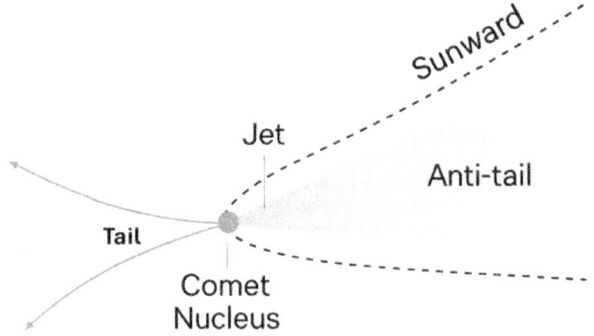

Figure 4. Schematic diagram of a comet showing narrow jets emerging from the nucleus and a broad sunward anti-tail formed by dust particles pushed by solar radiation.

Cluster 6 — Dust morphology

13. **Strong / negative polarization.** Polarimetric measurements (tracking how light's orientation changes upon reflection or scattering) show unusually high — and in some cases negative — polarization. Negative polarization means the light waves are lining up in the *opposite* direction from what we expect. It's like shining a flashlight into fog and seeing the reflected light twist in a direction fog normally doesn't produce.
  Tension: This implies unusual dust grain properties (size, shape, or composition).
14. **Sunward dust structure ("anti-tail").** At certain angles, 3I/ATLAS displayed a dust feature pointing toward the Sun.
  Tension: Sunward anti-tails are possible but require specific dust dynamics and viewing geometry.

12

Cluster 7 — Arrival context

15. Proximity to the Wow! signal region: According to calculations by Prof. Loeb, 3I/ATLAS arrives from a sky region not far from the direction associated with the 1977 Wow! radio signal.
Tension: This is most likely a coincidence, but it is an eerie one given the cultural weight of the Wow! event.

Another key feature of 3I/ATLAS is its speed, though not classified as an anomaly it is still worth mentioning. With a striking estimated velocity of 58 km/s, it is — so far —the fastest object ever detected.

To understand what these anomalies collectively imply, we now turn to a tool designed for exactly this kind of situation: probabilistic reasoning.

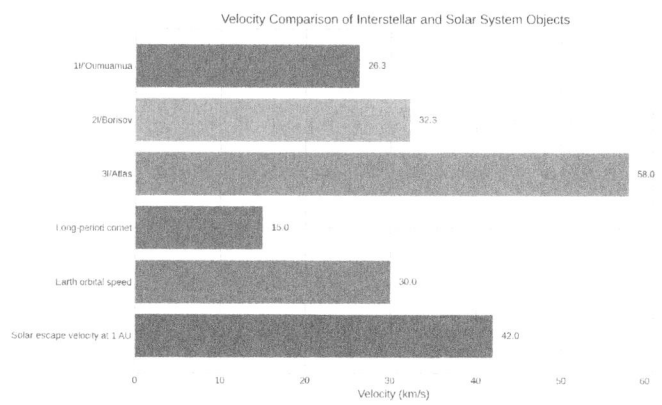

Figure 5. Interstellar objects such as 1I/'Oumuamua, 2I/Borisov, and 3I/ATLAS arrive with significantly higher inbound speeds than long-period comets, reflecting their unbound, interstellar origin. This chart compares these objects earth's orbital speed and the solar escape velocity at 1 AU (i.e., the velocity earth would need to escape from the solar system).

Natural or artificial? A probabilistic analysis of 3I/ATLAS' observations

When scientists encounter anomalies as puzzling as 3I/ATLAS', they often reach for a quiet but powerful tool: probabilistic thinking. This approach doesn't deliver certainty. Instead, it offers a disciplined way of asking which explanations require the fewest unlikely coincidences. Bayesian reasoning doesn't tell us what to believe—it tells us how much to adjust our beliefs as new evidence arrives—

The idea is straightforward. Instead of asking, "What is 3I/ATLAS?" we ask a more productive question:

"Given what we see, which explanation has to work harder?"

This is the same logic we use in everyday life. If you flip a coin a hundred times, you expect roughly half heads and half tails. Nothing strange there. But if you flip a hundred times and get ninety heads, you don't prove the coin is rigged — you simply start to wonder whether "fair coin" is still the best explanation. The more unusual the pattern, the more the balance shifts toward the idea that something other than chance might be involved.

3I/ATLAS presents a similar situation. None of its features are impossible to explain naturally. But many of them are unusual, and they don't all point in the same direction. Some relate to chemistry, some to geometry, some to dynamics, some to jet behavior. Each one, taken alone, is a "that's odd" moment. All seven together start to look like the astrophysical equivalent of 'my dog ate my homework'—technically possible, but increasingly hard to say with a straight face.

To make this intuitive, imagine labeling each unusual feature under a purely natural explanation:

- **Easy to explain.**

- **A bit stretched**

- **Really strained**

If 3I/ATLAS had one or two "a bit stretched" features, the natural explanation would remain comfortably ahead. But it has many such features — and a few that fall squarely into the "really strained" category, such as the nongravitational push without visible breakup, the narrow and stable jets, and the precise alignment with Jupiter's Hill sphere.

Now imagine — purely as a toy model — that each unusual feature has *rare* chance of occurring naturally . One such feature is no problem. Two are still fine. But stack five or six of them, and the odds of all of them lining up by accident start to look like:

rare × *rare* × *rare* × *rare*... which quickly becomes *very rare*.

This doesn't prove anything. But it does shift the balance of plausibility.

Under a **non-natural explanation** — whether a passive observer, a test platform, or a cautious scout — many of these features become easy to explain. The retrograde ecliptic path looks targeted. The stable jets look like controlled micro-maneuvering. The nongravitational acceleration looks like low-signature thrust. The Jupiter alignment looks like a deliberate gravitational waypoint. Again, none of this proves anything. It simply means that under a controlled-object model, fewer features require special pleading. But that combination does require one enormous assumption: that somebody made it. That single leap — not the physics — is what makes scientists cringe.

The **natural explanation** remains the default by scientists. Nature is diverse, and our sample of interstellar objects is tiny. But explaining 3I/ATLAS as a natural object requires a patchwork of exceptions: unusual chemistry, unusual dust, unusual jets, unusual trajectory, unusual acceleration. Each one is "rare but possible." Together, they stretch the natural scenario into a mosaic of coincidences.

Probabilistic thinking gives us clarity rather than certainty — a sense of which explanations flow naturally from the data and which ones require stacking unlikely events. It doesn't tell us what 3I/ATLAS is. It tells us how hard each explanation has to work.

And in science — as in life — the explanation that works the least hard is usually the one worth watching most closely.

A super quick view of Bayesian statistics (without the math)

When scientists talk about "Bayesian reasoning," they're not invoking anything exotic. It's simply a structured way of updating your beliefs as new information arrives. You start with an initial assumption — a *prior* — and then you adjust that assumption as evidence accumulates. For instance:

Imagine you're walking through a forest, and you hear a rustling sound behind a tree. Your prior belief is that it's probably something ordinary — a squirrel, a bird, the wind. Most rustling sounds in forests come from ordinary things.

Then you hear a low growl.

Your brain updates. The "ordinary explanation" is still possible, but the new evidence nudges you toward a different interpretation. You don't need certainty; you just need to adjust your expectations.

Then you see a large shadow move.

Your brain updates again. The "squirrel hypothesis" is now struggling. Now you have a *posterior* probability that it is not a squirrel. You don't need to know exactly what the creature is — you just know your earlier assumption is no longer the most plausible one. Obviously, you don't need to be certain of what is making the noise; you'd probably just run, hide, investigate, prepare to defend.

And this is the most important part: **Bayesian reasoning is not a tool for achieving certainty; it is a tool for <u>making decisions</u> when certainty is impossible.**

The same logic applies to 3I/ATLAS. We are not trying to prove that the object is artificial; we are trying to understand how much the anomaly pattern should shift our expectations, and whether that shift is large enough to justify paying attention. Bayesian reasoning helps us navigate that boundary:

the point where "probably nothing" becomes "worth preparing for," even if the underlying uncertainty never fully disappears.

This is Bayesian thinking in a nutshell:

- Start with a **reasonable** assumption.
- Let each new piece of **evidence** nudge you.
- Don't cling to your starting point if the evidence keeps piling up.

It's not about proving anything with absolute certainty. It's about being honest with yourself about how much the evidence should move you.

Figure 6. Bayesian inference is illustrated through a forest encounter. Panel 1: a rustling bush triggers a prior belief — likely a squirrel. Panel 2: a growl introduces new evidence. Panel 3: the posterior belief shifts toward a bear. Bayesian reasoning updates belief by integrating prior expectations with incoming data.

Bayes' rule says that the *posterior* probability of something being true or false *after* you see the evidence depends on two things:

1. How strongly you believed it before (your **prior**)
2. How surprising the **evidence** would be if it were true versus if it were false (the **likelihoods**)

In this context, it is important to remember these keywords: **probability** and **likelihood**.

Probability is how plausible a hypothesis seems before or after the evidence:

- *It's probably a squirrel* → **prior** probability.
- *It's probably a bear* → **posterior** probability.

Likelihood how well a hypothesis explains the evidence.

- *If it were a squirrel, how likely how likely are a growl and a large shadow? Not very likely.*

If the evidence is exactly what you'd expect under your hypothesis, your confidence goes up. But, if the evidence is surprising under your hypothesis, your confidence goes down. That's all Bayesian reasoning is: **Start somewhere reasonable. Then let the evidence push you around.**

Bayesian thinking is used in:

- **Medicine:** Doctors update diagnoses as new symptoms or test results appear.
- **Astronomy:** Researchers update the probability that a signal is a real exoplanet vs. noise.

- **Particle physics:** Experiments update the probability that a new particle exists as data accumulates.
- **Climate science:** Models update predictions as new temperature and CO_2 data come in.
- **Machine learning:** Bayesian updating is the backbone of many modern algorithms.
- **Geophysics:** interpreting uncertain subsurface signals (e.g., seismic) and making decisions .

Bayesian reasoning shines when:

- Evidence is incomplete.
- Signals are noisy.
- Multiple explanations are possible.
- No single observation is decisive.
- You need to weigh many small clues together.

Most people think science works by "proving" things. It doesn't. Science works by updating your beliefs:

- You start with a reasonable assumption.
- You gather evidence.
- You adjust.
- You keep adjusting as more evidence arrives.

This is exactly the situation with 3I/ATLAS. We don't have definitive proof of anything — no radio message, no high-resolution image. What we do have is a stack of unusual features, each individually explainable but collectively difficult to dismiss. Bayesian reasoning gives us a disciplined way to ask: **How much should this full pattern shift our expectations?**

When applied to 3I/ATLAS, it helps us avoid two traps:

- **The dismissive trap:** "It's obviously just a comet."

- **The sensationalist trap:** "It must be aliens."

Bayes gives us a middle path: **Given what we see, which explanation has to work harder?**

Bayesian analysis applied to 3I/ATLAS

In this section, only a summary of the conclusions and insights is offered. For the mathematically inclined — the geeks, the purists, and the brave — the full derivation is waiting in Appendix 1. What matters here is the logic and the learnings more than the calculations.

The analysis compares two broad possibilities:

- **Natural:** 3I/ATLAS is a strange but ultimately natural interstellar object.

- **Non-natural:** 3I/ATLAS is, in some sense, controlled or engineered.

Most scientists begin with a strong assumption that interstellar objects are overwhelmingly natural. That's a sensible starting point. We have never confirmed an artificial one, and nature is very good at producing variety.

Step 1: Prior probability

When choosing numbers for our priors, it helps to acknowledge something fundamental about uncertainty. Former U.S. Secretary of Defense Donald Rumsfeld once captured this with surprising clarity:

"There are known knowns —things we know that we know— There are also known unknowns —things we know we do not know — But there are also unknown unknowns — the ones we don't know we don't know."

This is exactly the situation we face when assigning a prior probability to the idea that an interstellar visitor could be artificial. We are navigating all three layers at once:

- **Known knowns:** Natural interstellar objects exist — we have observed three so far.
- **Known unknowns:** We do not know the full range of possible interstellar object configurations; we can only compare them to known comets, asteroids, and dust-rich bodies. We do not know the technological landscape of the galaxy, the frequency of other civilizations, the likelihood of reconnaissance probes, or whether such objects have passed through our solar system before.
- **Unknown unknowns:** These speak for themselves. Perhaps Earth hosts the only life in the universe. Perhaps life is abundant. Perhaps the universe is a simulation. At this level, the space of possibilities is unbounded.

Because of this layered uncertainty, pretending that a single *prior* captures all reasonable starting assumptions would be misleading. Instead, we evaluate three different priors, each representing a different stance on how plausible artificial probes might be:

- **Skeptical case:** Assume a 0.1% probability that any given interstellar visitor is artificial, and a **99.9% probability it is natural**. That corresponds to odds of 999 to 1 in favor of natural (or 1 in 1000 of being non-natural).
- **Moderate case:** Assume a 1% probability that an interstellar visitor is artificial and **99% that it is natural**. That's 99 to 1 in favor of natural (or 1 in 100 of being non-natural).
- **Adventurous case:** Assume a 5% chance of artificial origin and **95% probability it is natural**. That's 19 to 1 (same as 95 to 5) in favor of natural (or 5 in 100 of being non-natural).

This approach doesn't hedge. It simply acknowledges reality: **our uncertainty has uncertainty.**

A keen reader would wonder, rightly, why the "skeptical" prior stops at one in a thousand instead of plunging to one in a million?. If skepticism is good, isn't skepticism better? In principle, yes — but only up to the point where the prior stops being cautious and starts behaving like a restraining order against learning.

A good way of thinking about our prior is this: if 1000 *interstellar objects* cross our neighborhood in a way we can detect, 999 are rocks and 1 could be artificial. This is very different from "1 in 1000 comets overall," which would be absurd.

If we were to set up a prior of **0.000001**, for instance (**a 99.99999% probability of being natural**), we would be saying that before seeing any data, we have already decided that this kind of event is so implausible that no ordinary evidence could ever change our minds. That's not skepticism; that's dogma.

By contrast, a prior of 1/1000 is still harsh. It says, "I doubt this very much," not "I have welded the door shut." It forces the artificial hypothesis to work hard, but it doesn't make the job physically impossible. Readers who prefer even more extreme priors are welcome to imagine them — just be aware that at some point you're no longer updating beliefs; you're simply watching the evidence bounce off a brick wall you built yourself.

Step 2: Looking at the evidence.

We clustered the anomalies so they wouldn't form a union.

Now we turn to the evidence. 3I/ATLAS isn't strange in one way — it's strange in many ways. We outlined roughly fourteen anomalies earlier, though many share common physical causes. To keep the analysis disciplined and avoid *double dipping*, we compress them into the seven clusters already introduced, each representing a distinct scientific domain:

1. **Orbital plane and direction (likelihood ~3:1):** Retrograde and near-ecliptic is possible for a natural object, but not common. A controlled object might

choose this path deliberately. *A likelihood of 3 to 1 gives this a small push toward non-natural.*

2. **Jupiter Hill sphere alignment (likelihood ~10:1):** A near-perfect pass by Jupiter's gravitational boundary is strikingly unlikely by chance. For a guided object, using Jupiter as a waypoint or gravity-assist target, it makes perfect sense. *A likelihood of 10 to 1 gives this a Strong push toward non-natural.*

3. **Composition, nickel-rich, low water, CO_2-heavy, unusual dust (likelihood ~1:1):** Unusual, but not impossible. Easy to explain both under a natural and an engineered scenario. *A likelihood of 1 to 1 keeps this Neutral.*

4. **Photometric behavior, rapid brightening, blue tint, delayed activation (likelihood ~1:1):** Natural comets can do this, but it's not typical. We also lack a large sample of interstellar objects. *Neutral.*

5. **Jets, stability, and non-gravitational acceleration (likelihood ~10:1):** Strong, stable jets and smooth acceleration without breakup are very difficult to explain for natural models and easy for controlled propulsion. *A likelihood of 10 to 1 gives this a very strong push toward non-natural.*

6. **Dust morphology, sunward "anti-tail", negative polarization (likelihood ~1:1):** Possible naturally, possible artificially. *Neutral.*

7. **Arrival direction near the Wow! region (likelihood ~1:1):** It is difficult to justify this as artificial, without entangling the probability to the

> Wow! Signal being non-natural. We treat it as a coincidence. *Neutral*

These likelihood numbers aren't just guesses — they come from a simple question scientists ask whenever they compare explanations: **"If this object were natural, how surprising would this feature be? And if it were non-natural, how surprising would it be?"**

For each cluster, we considered:

1. **How often does nature produce this behavior** (based on known comets, interstellar objects, and physical models)

2. **How naturally the same behavior would arise under control** (based on what a low-thrust, reconnaissance, or engineered object would reasonably do)

3. **How many special assumptions each explanation needs** (fine-tuned jets, rare compositions, precise coincidences, etc.)

Then we assigned conservative likelihood ratios — *always erring on the side of the natural explanation to avoid forcing a conclusion.*

For example:

- A retrograde, near-ecliptic path is **uncommon** for natural objects but **expected** for a survey trajectory → about **3:1** in favor of control.

- A near-perfect pass by Jupiter's Hill sphere is **very unlikely** by chance but **routine** for a guided object → about **10:1**.

- Strong, stable jets and smooth acceleration without breakup are **hard** for natural models but **easy** for controlled propulsion → another **10:1**.

Other clusters don't strongly favor either explanation, so they get **1:1**.

When we multiply these likelihoods — 3 × 10 × 1 × 1 × 10 × 1 × 1 — the overall pattern favors the non-natural explanation by about **300 to 1 likelihood**. This only means the evidence fits that explanation far easier than the purely natural one.

Now let's see how this changes our prior probabilities.

Step 3: Putting it all together.

None of these features alone is decisive. But together, they form a pattern that is difficult to dismiss as coincidence. A natural explanation can account for each anomaly individually — but only by invoking a different "rare but possible" condition for almost every feature.

A non-natural explanation doesn't need to stretch as far. Many of the behaviors we've discussed — the trajectory, the jets, the acceleration, stability, look like things you might reasonably expect from a controlled object. When we treat these anomalies as evidence and update our assumptions accordingly, the numbers shift dramatically.

Skeptical case: If you begin with a **prior** of only a **0.1% probability** that 3I/ATLAS is artificial — a one-in-a-thousand starting point — after the Bayesian update (see Appendix 1 for the full calculation) the anomaly stack pushes the **posterior** that all the way up to roughly:

- **23% probability of being non-natural.**
- **77% probability of being natural.**

In other words, the data alone is strong enough to move our initial prior significantly (from 0.1% to 23%). Twenty-three percent may look small, but it is still significant. A useful way to think about it is with a weather analogy: *if you see a 23% chance of a storm, you don't cancel your plans, but you probably bring an umbrella.* Or in medical terms*: a 23% chance of a condition is not a diagnosis, but it is absolutely worth further testing.*

Moderate case: If you begin with a **prior** of **1% probability** that an interstellar visitor could be artificial — still a very conservative stance — the updated **posterior** probability rises to about:

- **75% probability of being non-natural.**
- **25% probability of being natural.**

The natural explanation becomes the underdog.

Adventurous case: If you start with a **5% prior** — still far from assuming anything extraordinary — the updated **posterior** probability becomes:

- **94% probability of being non-natural.**

- **6% probability of being natural.**

At this point, the natural explanation is struggling to keep up with the evidence.

From this point forward, our base case will be the **skeptical result** described above. This isn't a confession of personal belief; it's a deliberate methodological choice. Amongst the priors we considered, this one gives the smallest imaginable starting probability that anything unusual is happening — it is the prior most generous to coincidence, noise, and the ordinary churn of randomness.

A good way of thinking about this argument is: *If the evidence can move even this stubborn prior, then it would move any of the gentler ones even more.* The moderate and adventurous priors still matter, but mainly as sensitivity checks. They show how different temperaments of belief would update when confronted with the same data.

By fixing the skeptical prior as our baseline, we ensure that every comparison is measured against the most demanding starting point. *Think of it as asking the harshest possible critic to review the evidence first. If we start from the most pessimistic scenario and the evidence still pushes us into non-natural territory, that alone makes the case interesting.*

Step 4: Sensitivity analysis

Bayesian reasoning is often described as a mathematical tool, but its power is easier to grasp when **visualized** rather than calculated. Any probabilistic analysis requires assumptions about *priors* and *likelihoods*, and a careful reader will naturally ask what happens if those assumptions change. Another scientist could choose different values and arrive at a different numerical answer. That concern is valid — and it is exactly why visualization matters. To illustrate this sensitivity, the next three figures provide a compact visual summary of how priors, likelihood ratios, and posteriors interact. They are not meant to "prove" anything; they show how belief moves when confronted with structured evidence. A full explanation of how these graphics are generated is provided in Appendix 1.

The first figure shows how the posterior probability changes as a function of the prior when the strength of the evidence is held constant. Even extremely skeptical priors — fractions of a percent — rise sharply once the likelihood ratio reaches values consistent with the anomaly stack. A conventional 5% significance line is included not as a decision rule, but as a familiar reference point: the moment when a hypothesis stops being negligible and starts demanding attention. This should not be interpreted as a frequentist p value (the point at which a result could be explained simply by chance) or as evidence exceeding a 5% significance threshold. We include it simply because it provides readers with a recognizable landmark — a visual cue that anchors the scale of the plot. It marks the point where a probability stops feeling negligible and starts feeling worth noticing, even though it plays no formal role in Bayesian inference. In that sense, it functions as a psychological guidepost rather than a threshold of belief, helping general audiences appreciate how quickly the posterior rises once the evidence is applied.

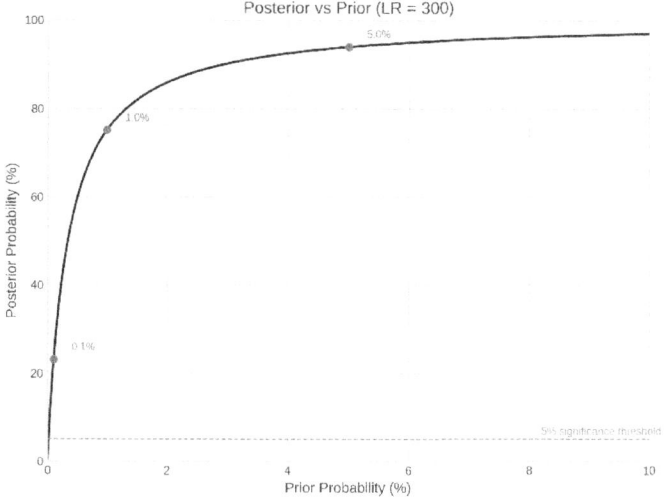

Figure 7. The continuous line shows how the posterior probability changes as we vary the prior. The three discrete cases used earlier in the chapter are shown as red dots. Even with conservative priors, the posterior rises sharply given the strength of the evidence.

The second figure widens the lens by showing how the posterior responds to different strengths of evidence. Three curves illustrate weak, moderate, and strong likelihood ratios. All of them rise steeply, and all cross the psychological 5% line for even modest priors. This is the essence of the argument: the posterior is not fragile, and it does not depend on numerical fine-tuning. A broad range of priors and likelihood ratios lead to the same **qualitative** conclusion.

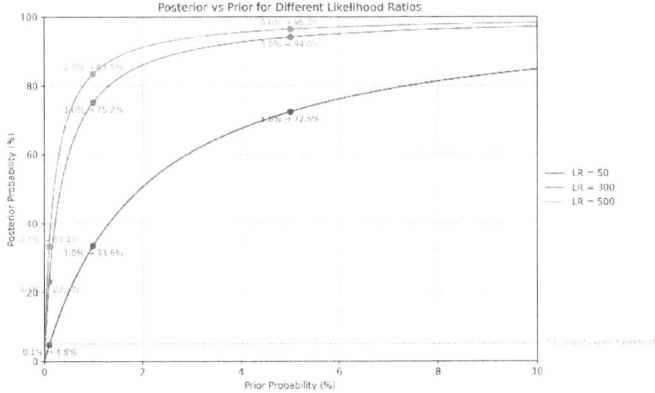

Figure 8. The three continuous lines show how the posterior probability changes when we vary the likelihood ratio — how strongly the evidence pushes us toward the non-natural hypothesis. Even with a very conservative weight (blue line), the posterior rises sharply above the 5% significance line. The markers show discrete prior-to-posterior calculations.

The third figure maps the entire Bayesian landscape at once. The horizontal axis is the prior, the vertical axis is the posterior, and the color encodes the likelihood ratio required to move from one to the other. Contour lines mark likelihood ratios of 10, 20, 50, 100, 300, and 500, (our base case uses 300:1 as likelihood) and a dashed line highlights the 5% psychological threshold where a hypothesis stops being negligible. Two features stand out immediately: (1) the region where the posterior exceeds 5% is very large, and (2) only extremely tiny priors combined with extremely weak evidence keep the posterior below that line. The update is robust, not brittle.

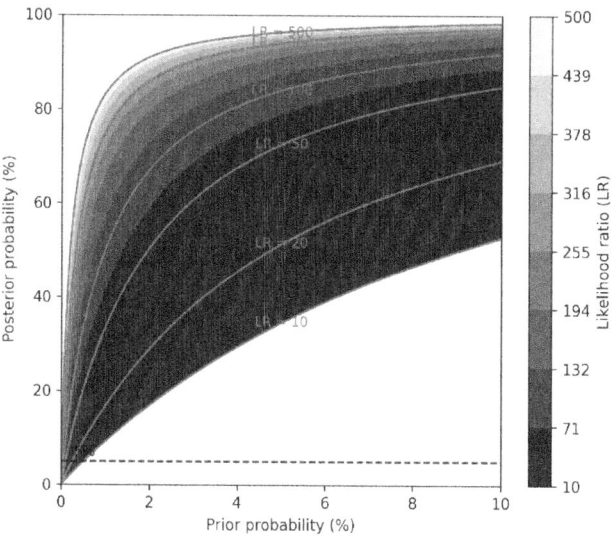

Figure 9. This plot shows how the posterior probability (vertical axis) responds to different combinations of priors (horizontal axis) and likelihood ratios (color bands). Regardless of which combination is chosen, the posterior crosses the 5% significance threshold across a wide region of the parameter space.

Taken together, these figures show that the anomaly pattern does not require extreme assumptions to shift our expectations. The Bayesian update is not driven by any single feature of 3I/ATLAS, but by the cumulative weight of many small tensions pointing in the same direction. With this visual foundation in place, we can now turn to the central question: what do these results mean?

33

What these results mean

As stressed before, these results don't prove that 3I/ATLAS is artificial. What they lead us to consider is something more subtle — and arguably more important.

If you start with a skeptical prior, the full anomaly stack of 3I/ATLAS still pushes you strongly toward taking the non-natural explanation seriously. The natural explanation remains possible, but it requires stacking many unlikely coincidences. The non-natural explanation requires fewer stretches. It becomes a serious competitor, not a fringe idea.

To put the 23% posterior in context: in frequentist statistics, researchers often reserve the word "significant" for patterns that would arise by random chance less than 5% of the time (the familiar $p < 0.05$ threshold). This 5% line is not a law of nature, but a long-standing psychological and methodological convention for controlling Type I error rates — essentially a rule of thumb that says: "If this pattern would appear by accident fewer than one time in twenty, we should at least pay attention

The reason this matters is that the 23% emerges even under our most skeptical prior. In other words, the data are pushing back against a starting assumption that is already leaning heavily toward "nothing to see here." It's not enough to declare victory, but it is enough to say, **"Something is pulling the needle."**

Table 1. Welcome to 23%: where unlikely things still happen

What does a 23% probability *feel* like?

Event	Approx. Probability	Why it's a good analogy
A 23% chance of rain	~20–25%	You don't cancel the picnic... but you keep an umbrella nearby.
Getting two heads in a row	25%	Not the most common outcome, but nobody is surprised when it happens.
A basketball player making a 3-pointer	~25% for an average shooter	You wouldn't bet on it every time, but it goes in often enough to matter.
Picking a movie with a nonsensical plot twist	~20–30%	Streaming roulette is a dangerous game.
Seeing a squirrel on a short walk in a city park	~20–30%	Not guaranteed, but you'd never bet against it.
Having at least one cracked corner on your phone screen	~25% of users	A perfect metaphor for "not typical, but absolutely not rare."
Talking to yourself out loud	~25% of people admit it	And the other 75% are probably doing it too.
Getting the "golden fry" in your order	~20–25%	Rare enough to feel special, common enough to expect occasionally.

As mentioned before, Bayesian reasoning doesn't aim to deliver certainty; it helps us make decisions when information is scarce and imperfect. And in the case of 3I/ATLAS, the evidence shifts our expectations enough that the "just a weird comet" story is no longer overwhelmingly dominant. That leaves us with an unavoidable question — if the natural explanation is no longer the default by sheer weight of probability, **what do we do now?**

Implications of a non-natural possibility

The Bayesian analysis tells us the non-natural explanation is strong enough that it deserves to be treated as a legitimate scientific possibility. Not the default, not the conclusion — but a contender that rational observers should keep on the table.

Once we accept that the artificial scenario is no longer fringe, the next question becomes: **If 3I/ATLAS were non-natural, what kind of object might it be?**

If we are willing to entertain even the remote possibility that 3I/ATLAS could be artificial, we need a framework for thinking about the unknown. The object's behavior does not force a single interpretation — it opens a landscape of possibilities. Some of those possibilities are mundane, some are unsettling, and some are simply unfamiliar because we have never had to think about them before. A disciplined response begins not with certainty, but with structure.

If an interstellar visitor behaves in ways that stretch our natural models, there are at least four broad categories of explanation available to us. These categories are not conclusions; they are lenses. Each one captures a different kind of intent — or non-intent — and each one makes different predictions about how an object like 3I/ATLAS should behave. These scenarios are not claims about extraterrestrial motives. They are behavioral categories — the same analytical tools we would use to interpret any unknown agent, human or non-human.

The possibilities are:

1. **Preparatory tests for first contact.** A sequence of small, escalating anomalies designed to measure our awareness, our stability, and our interpretive discipline before any direct communication.

2. **Passive observation.** A long-term, non-intrusive watcher that studies us the way we study ecosystems — quietly, without interference, and without revealing its own motives.

3. **A preamble to a threat.** A reconnaissance pattern optimized not for communication or curiosity, but for mapping vulnerabilities, blind spots, and response behaviors.

4. **A purely natural phenomenon.** An object whose strangeness reflects the limits of our models rather than the presence of intent — a reminder that the universe is more diverse than our expectations.

These four scenarios define the interpretive space in which 3I/ATLAS can be understood. They are the coherent categories that span the spectrum of possibilities, from benign to indifferent, to dangerous, and to entirely natural. A mature civilization would evaluate all four with equal discipline, resisting the temptation to collapse prematurely into comfort or fear.

The goal is not to decide which scenario is true. The goal is to understand how each scenario interprets the same evidence — and what each one would imply about the behavior we observed. With this framework in place, we can now examine how 3I/ATLAS fits into each possibility, and what its anomaly stack reveals when viewed through these four distinct lenses.

The scenarios do not compete for truth. They compete for explanatory power.

Scenario 1: Preparatory test for first contact

In this scenario, 3I/ATLAS would be part of a deliberate, peaceful sequence. Many civilizations — including our own — would send reconnaissance probes long before attempting direct communication. A small, maneuverable, data-gathering object is the safest and most efficient way to learn about a planetary system: its chemistry, its energy use, its technological signatures, and whether its inhabitants are ready for contact.

If 3I/ATLAS were such a probe, several of its most unusual features fall naturally into place. A controlled trajectory through the ecliptic would be the obvious choice for an intelligence seeking a panoramic survey of the planetary system. A precise pass near Jupiter's Hill sphere would be consistent with gravitational way-pointing or a mapping maneuver. A brief observational window near the inner Solar System, followed by a clean departure, would fit a mission designed to gather information without revealing intent.

Nothing about this scenario requires urgency or threat. It frames 3I/ATLAS as a quiet prelude to a much longer, cautious process.

In such a scenario, we would not expect overt communication, dramatic maneuvers, or anything that could be interpreted as a declaration of intent. Instead, we would expect something quieter and more methodical: a sequence of behaviors designed to probe our detection capabilities, map our observational blind spots, and assess how we interpret ambiguous data. In this frame, 3I/ATLAS is not trying to talk to us — it is trying to understand **how we see**.

Viewed through this lens, the object's retrograde ecliptic alignment becomes a deliberate choice for maximizing observational coverage. The Jupiter alignment becomes a purposeful waypoint rather than a coincidence. And the most striking cluster — the stable jets and smooth nongravitational acceleration— can be interpreted as controlled micro-thrusting: small, deliberate perturbations that test how we reconcile conflicting data about motion, mass, and structural integrity.

Other features — such as the dust morphology, composition, and photometric behavior — remain fully compatible with natural explanations, but in this scenario, they serve a different role. They become part of a broader pattern of "edge-case" signals: not impossible, not definitive, but just unusual enough to reveal how quickly we notice deviations from our expectations. In this interpretation, 3I/ATLAS behaves like a reconnaissance instrument designed to learn about us indirectly — not by announcing itself, but by watching how we respond to something that sits just outside the boundaries of the familiar.

Scenario 2: Passive observation

In this scenario, 3I/ATLAS is not preparing for contact at all. It is simply a scientific instrument — an interstellar weather satellite — sent to gather data on many systems, not ours specifically. Its trajectory would be optimized for efficiency rather than interaction, and its behavior would reflect a mission designed to observe, sample, and move on. In this frame, 3I/ATLAS is indifferent to us. We are not the target; we are just one of many stops along a long survey route.

This scenario fits well with the idea of a civilization that seeds the galaxy with autonomous probes to map environments, track long-term changes, or study planetary evolution. It is the most "boring" artificial scenario — and perhaps the most plausible.

If 3I/ATLAS is not preparing for contact but simply watching, its behavior would look different from a test platform. A passive observer does not probe, calibrate, or challenge our detection systems. It does not introduce controlled perturbations or attempt to map our interpretive limits. Instead, it behaves like a long-baseline survey instrument: quiet, stable, and unobtrusive. Its goal is not to interact with us, but to gather information about the Solar System as it passes through.

Seen through this lens, several of 3I/ATLAS' strongest anomalies fall naturally into place. The retrograde ecliptic alignment becomes a straightforward observational geometry. An object arriving along the plane of the planets — but moving opposite their direction — would enjoy a sweeping, high-contrast view of the Solar System's architecture. This is the same geometry we would choose for a reconnaissance fly-through of an unfamiliar planetary system.

The Jupiter Hill sphere alignment, one of the most statistically striking features, also fits comfortably here. A passive observer passing through a planetary system might use large gravitational bodies as reference points, much as our own spacecraft use planetary flybys to refine trajectories or stabilize long-baseline observations. The precision of 3I/ATLAS' projected encounter with Jupiter's gravitational boundary could be interpreted as a navigational convenience rather than a deliberate test — a way to maintain a stable path while minimizing fuel or energy expenditure.

Other features — such as the dust morphology, composition, and photometric behavior — do not strongly favor either hypothesis in the Bayesian analysis. In this scenario, they are descriptive rather than diagnostic. They tell us something about the object's material history or construction, not its intentions. A probe designed for long-duration travel might use alloys or volatiles that differ from typical cometary compositions. A natural object formed in an unusual environment might show the same signatures. In a passive-observation scenario, these features are clues to origin, not signals of behavior.

Even the jets, narrow outflows, and smooth nongravitational acceleration — the most challenging cluster for natural models — can be interpreted without invoking intent. A passive observer might employ controlled venting or engineered thermal pathways to maintain stability during perihelion. A natural object with unusual internal structure might behave similarly. In this scenario, the question is not "What is it trying to tell us?" but "What does this behavior reveal about its construction or composition?"

In this scenario, 3I/ATLAS is not evaluating us. It is not hiding, signaling, or probing. It is simply passing through, gathering data as it goes, following a trajectory that optimizes its own observational goals. Its silence is not strategic — it is intrinsic. A passive observer has no reason to respond to our signals, no incentive to reveal itself, and no need to alter its course in reaction to our presence. Where the first scenario frames 3I/ATLAS as an examiner, this one frames it as a traveler with a camera — a quiet instrument drifting through the Solar System, recording what it sees, indifferent to whether we notice.

Scenario 3: Threat preamble

This is an uncomfortable scenario, but intellectual honesty requires considering all possibilities. In this frame, 3I/ATLAS would be performing reconnaissance for strategic rather than scientific reasons. A controlled trajectory, close planetary passes, and detailed mapping could be interpreted as information gathering for future action. Nothing in the data demands this interpretation, and nothing in 3I/ATLAS's behavior suggests hostility. But if one entertains the artificial hypothesis at all, this scenario must be included for completeness. It represents the branch of the decision tree where reconnaissance is not a prelude to contact, but to assessment.

If 3I/ATLAS represents the early stages of a threat — not an attack, but a pre-engagement maneuver — its behavior would look neither theatrical nor overtly hostile. A civilization capable of interstellar travel would not need to announce itself with spectacle. Instead, a threat preamble would be defined by strategic positioning, information gathering, and risk minimization.

The goal would be to enter the Solar System quietly, map the environment, assess defenses, and approach key gravitational or observational vantage points without triggering panic or retaliation.

Seen through this lens, several of 3I/ATLAS's strongest anomalies take on a more tactical coherence. The retrograde ecliptic alignment becomes a deliberate choice: a path that maximizes visibility of the planetary system while minimizing the likelihood of collision or interception. Approaching along the plane of the planets —but against their direction— provides a sweeping, high-resolution survey of orbital traffic, energy signatures, and technological infrastructure. It is the same geometry a reconnaissance craft might choose when entering a potentially hostile system.

A further detail becomes significant in this frame: **3I/ATLAS reached perihelion while positioned almost exactly opposite the Sun from Earth's point of view**, effectively placing itself in a solar eclipse relative to us. This geometry is observationally inconvenient for Earth-based telescopes and would be an ideal moment for an object to minimize scrutiny during its closest solar pass. Combined with the fact that its inbound trajectory "avoided" any close approach to Earth, the pattern resembles intentional concealment — the behavior of an intelligence aware of our presence and deliberately choosing to limit our observational opportunities.

Features such as the dust morphology, composition, and photometric behavior — which the Bayesian analysis treats as neutral — do not carry strategic meaning here. They may simply reflect construction materials or environmental history. In this scenario, they are neither signals nor attempts at concealment; they are incidental properties of an object whose purpose lies in its trajectory and behavior, not its surface chemistry.

The jets, narrow outflows, and smooth nongravitational acceleration, however, take on a different significance. These are the most challenging features for natural models and the ones that most strongly favor control. In a threat-preamble scenario, they can be read as low-signature propulsion: precise, stable adjustments that allow the object to refine its path without revealing a conventional engine.

A threat preamble does not require high thrust; it requires subtlety. The fact that the nucleus remains intact despite the implied mass loss may indicate internal reinforcement or engineered vent geometry — characteristics consistent with a craft designed for maneuverability rather than fragility.

The most consequential feature in this scenario is the Jupiter Hill-sphere alignment. A strategically minded intelligence would almost certainly exploit major gravitational bodies for positioning, shielding, or staging. Passing near Jupiter's gravitational boundary offers several advantages:

- It provides a stable vantage point for observing the inner Solar System.
- It allows the object to mask its trajectory behind Jupiter's bulk.
- It offers a low-energy pivot point for redirecting deeper into the system.
- It tests whether the system's inhabitants can track a precise gravitational encounter.

The precision of 3I/ATLAS's projected passage — especially following a measurable nongravitational acceleration — is exactly the kind of maneuver a cautious, strategically minded intelligence might execute before committing to deeper engagement.

In this scenario, 3I/ATLAS is not here to communicate or to test our scientific reflexes. It is here to position itself. Its silence is not restraint but indifference. Its trajectory is not a survey path but a pre-deployment arc.

Where the first scenario frames 3I/ATLAS as an examiner and the second as a traveler, this scenario frames it as a scout: a quiet, methodical precursor to something larger, mapping the terrain before the main force arrives.

Scenario 4: Natural phenomenon

In this scenario, 3I/ATLAS is simply a strange but natural interstellar visitor whose anomalies reflect the diversity of exoplanetary debris rather than any form of control or engineering. In this frame, the unusual trajectory and jets are not signals of intent — they are reminders that nature is more varied than our current catalog.

This scenario is science's base case. A 23% probability of artificial origin (under the most skeptical prior) still means a 77% probability that 3I/ATLAS is natural. Even under more generous priors, the natural explanation is never ruled out; it simply becomes less favored as the anomaly stack grows.

The natural-phenomenon scenario is the one that demands the most intellectual discipline. It asks us to assume that every unusual feature of the object, no matter how striking, emerges from the physics of its formation and the stresses of its interstellar journey.

In isolation, this is a reasonable position. Nature is diverse, and our sample of interstellar objects is tiny. But when we consider all of 3I/ATLAS's observed features together, the natural explanation becomes less a single clean story and more a patchwork of exceptions, edge cases, and special conditions.

From this perspective, the retrograde ecliptic alignment is simply a coincidence of celestial mechanics. Some interstellar objects will inevitably pass near the ecliptic plane, and some will arrive on retrograde paths.

But the combination — retrograde motion and near-perfect alignment with the Solar System's "tabletop" — is statistically uncommon. A natural explanation is possible, but it requires us to accept that 3I/ATLAS just happens to follow one of the more observationally convenient geometries.

Features such as the dust morphology, composition, and photometric behavior — which the Bayesian analysis treats as neutral — fit comfortably within this scenario. They each have natural explanations, and none of them strongly favor artificiality.

But they do point to different formation environments and different physical histories. None of these features break physics — but each one nudges the object toward the edge of what we consider typical.

The nongravitational acceleration and the intact nucleus present a deeper tension. To explain the observed acceleration naturally, we must assume significant mass loss.

To explain the intact nucleus, we must assume that the mass loss is either extremely uniform, extremely localized, or occurring in ways we cannot resolve. Again, none of this is impossible — but it requires a chain of assumptions that must all be true simultaneously.

The most challenging feature for the natural scenario is the Jupiter Hill sphere alignment. A natural object can pass near Jupiter's gravitational boundary by chance.

But the precision of 3I/ATLAS's projected encounter — especially after a measurable nongravitational acceleration — is statistically striking. A natural explanation is still possible, but it requires us to accept that 3I/ATLAS's trajectory is an unusually clean outlier in a dataset of only three known interstellar visitors.

In this scenario, 3I/ATLAS is a wanderer: a fragment of another planetary system carrying the chemical and structural fingerprints of its origin. Its features are not signals but consequences. Its trajectory is not strategic but serendipitous. Its silence is not meaningful; it is intrinsic.

But honesty requires acknowledging the limits of this interpretation. A natural explanation can account for each feature of 3I/ATLAS — but only by invoking a different exception, a different edge case, or a different special condition for nearly each one.

The scenario is viable, but not simple. It is a mosaic of "unusual but possible" explanations stacked on top of one another.

Where the other scenarios explore intention, this one explores complexity. It reminds us that nature is capable of surprising us

— but also that surprise, when repeated across many independent features, becomes its own kind of signal.

Comparative matrix

Now that we have four coherent scenarios — preparation for first contact, passive observer, threat preamble, and natural object — the next step is to compare how each one fits the actual anomaly structure of 3I/ATLAS. The goal is not to decide which scenario is correct, but to see how each one interprets the same evidence.

A comparative matrix helps clarify this. It shows, at a glance, how each scenario handles the seven anomaly clusters used in the Bayesian analysis. Some clusters carry strong discriminating power (orbital geometry, Jupiter alignment, jets). Others are neutral (composition, photometry, dust morphology, arrival direction). By mapping each scenario against these clusters, we can see where the interpretations converge, where they diverge, and where they rely on coincidence, intent, or engineering.

This matrix is not a scorecard. It is a clarity tool — a way to make explicit the assumptions each scenario must adopt to remain viable. It shows how the same data can support very different narratives, and how the weight of the evidence shifts depending on the interpretive frame.

Table 2. Comparison of how four possible scenarios fit the seven major observational features of 3I/ATLAS. Green check marks indicate features that align naturally with a scenario; yellow dashes indicate compatibility without strong support; red crosses indicate features that require coincidence or multiple special assumptions.

Feature	Scenario 1 First contact Preparation	Scenario 2 Passive Observer	Scenario 3 Threat preamble	Scenario 4 Natural Object
1. Orbital Geometry (retrograde, near-ecliptic)	✓ A good path for scanning the whole Solar System	✓ Fits efficient observational geometry	✓ Fits a tactical entry path	Coincidental but possible
2. Jupiter Hill sphere Alignment	✓ Looks like a planned gravity assist or mapping pass	✓ Fits a navigational convenience	✓ Fits a strategic positioning	X Low probability coincidence
3. Jets & Non-Gravitational Acceleration	✓ Fits controlled, gentle propulsion	Could reflect engineered stability or passive control	✓ Fits a low-signature propulsion	X Requires multiple fine-tuned natural assumptions
4. Composition (nickel-rich, low water)	Could be incidental materials	Incidental	Fits a functional material choice	✓ Well within natural compositional diversity
5. Photometric Behavior (brightening, blue tint)	Could come from engineered materials reacting to sunlight	Normal for some materials or surface types	Could be a byproduct of construction	Rare but natural
6. Dust Morphology (sunward anti-tail)	A normal dust-release pattern that tests how we interpret unusual shapes	Simply a byproduct of illumination and viewing angle	Not optimized for concealment; consistent with ordinary dust behavior	✓ A well-known natural effect caused by dust grains lagging behind the object
7. Arrival Direction (Wow! region proximity)	Coincidence	Coincidence	Coincidence	Coincidence

Taken together, the matrix makes one point unmistakably clear: the different scenarios do not compete on equal footing when it comes to explaining the strongest anomalies. The artificial scenarios — whether framed as preparation for contact, passive observation, or strategic reconnaissance — all provide clean, internally consistent interpretations of the orbital geometry, the Jupiter alignment, and the smooth nongravitational acceleration. These features fall naturally into place when intent or control is allowed into the model.

The natural scenario, by contrast, remains viable but increasingly strained: it must treat the same high-weight anomalies as independent coincidences, each requiring its own special conditions. Meanwhile, the neutral clusters — composition, photometry, dust morphology, and arrival direction — do little to discriminate among the possibilities, which is why they appear as yellow across the board.

The matrix therefore doesn't tell us which scenario is true, but it does reveal how the explanatory burden shifts. Artificial scenarios gain coherence as the anomaly stack grows; the natural scenario retains plausibility but at the cost of accumulating exceptions. The result is not a verdict, but a clarified landscape — a structured view of how each scenario fits the data, and where each one must stretch to remain credible. The artificial scenarios offer coherence at the cost of implication; the natural scenario offers safety at the cost of complexity. We are left standing between them, staring at a visitor that refuses to tell us what it is, or why it came, or whether it even noticed us at all.

But the most important thing is not whether 3I/ATLAS is natural or artificial, but what the updated probabilities imply for our behavior. Bayesian reasoning asks us to decide how seriously to take a possibility once the evidence has shifted. In the same way that a rustle→growl→shadow sequence in a forest doesn't prove a bear but still justifies stepping back, the anomaly pattern of 3I/ATLAS doesn't prove artificiality — but it does justify paying attention. The question is no longer "Is it natural or not?" but **"Given the updated odds, what is the rational posture to adopt as we move forward?"**

Probabilities alone don't tell us the whole story. The matrix clarifies only how each scenario fits the data — not why an artificial object, if it is one, would behave this way. The anomalies themselves don't just describe motion or composition; they hint at patterns of choice, restraint, and priority. A controlled trajectory suggests one kind of mindset, a silent fly-through another, and a cautious gravitational encounter still another. If 3I/ATLAS is natural, these patterns dissolve back into coincidence. But if it is artificial, then the behavior we observe is not just physics — it is psychology.

To understand what kind of intelligence might produce these signatures, and what its goals or constraints might be, we need to shift from orbital mechanics to behavioral archetypes.

The four scenarios give us a structured way to think about 3I/ATLAS — but they don't exist in a vacuum. No object, no signal, and no anomaly ever does. To understand how 3I/ATLAS fits into any of these possibilities, we need to widen the lens and look at the full sequence of events that led us here. Long before 3I/ATLAS appeared, the sky had already delivered a series of strange visitors, plus one unforgettable signal, each arriving with its own timing, its own character, and its own unanswered questions. Whether these events are connected or coincidental, they form the backdrop against which 3I/ATLAS must be interpreted.

So, before we analyze intent or behavior, we turn to the anomaly stack itself — the pattern of arrivals that may, in hindsight, reveal a story much larger than a single object.

Chapter 2: The anomaly stack

If there's a pattern out there, we'll find it. If there isn't, we'll still give it a try.

Before 3I/ATLAS ever appeared, the sky had already begun to whisper. Not loudly, not clearly, but in a cadence just strange enough to make hindsight feel uncomfortable. A single radio burst in 1977. A tumbling shard from the stars in 2017. A textbook comet in 2019. Each event arrived alone, wrapped in ambiguity, easy to dismiss as coincidence. But when placed side by side, they begin to resemble something else — a sequence, a rhythm, a pattern of arrivals that probes different parts of our awareness.

This chapter steps into that possibility, but with care. What follows is a narrative that dances between science and speculation, just an exercise in responsible imagination. The goal is not to assert intent, but to explore whether these anomalies, taken together, form a pattern worth noticing. Here, we link a string of curious events — the 1977 Wow! radio burst, 1I/'Oumuamua, 2I/Borisov, and 3I/ATLAS — as a loose sequence of tests probing different aspects of our awareness. The idea began as a simple thought experiment: after noticing the cadence and character of these events, **I wondered what an external civilization might do if it were preparing first contact**, whether for study or for reasons we cannot yet guess.

History offers a sobering parallel. When civilizations meet for the first time, the encounter is often disastrous for the less advanced party. A cautious intelligence — one seeking to avoid catastrophe, not cause it — might proceed step-by-step, much as planners do when assessing an unfamiliar environment: reconnaissance, visual confirmation, baseline sampling, long-term pattern analysis. This analogy is meant to illuminate how an observer might gather and interpret signals, not to prescribe tactics or declare intent. It is a lens, not a conclusion. Imagine standing at the edge of a vast ocean, watching ripples that might be signals from another shore. Could these anomalies be breadcrumbs left by an intelligence beyond Earth? Or are they simply coincidences that reveal more about our pattern-seeking minds than about the universe itself? This chapter explores that tension — with curiosity, with caution, and with scientific humility.

Stage 1, Reconnaissance: are we listening? Wow! signal

Ohio State University's radio telescope – called Big Ear - was an odd-looking giant in the Ohio cornfields, a fixed radio telescope built from a vast sheet of aluminum ground and two enormous reflectors, one flat and one curved, facing each other across the field like silent walls. It couldn't swivel or tilt; instead, it let the sky drift overhead as Earth turned, sweeping a thin celestial strip night after night with monastic patience. For more than two decades it listened this way, running one of the longest continuous SETI searches ever attempted.

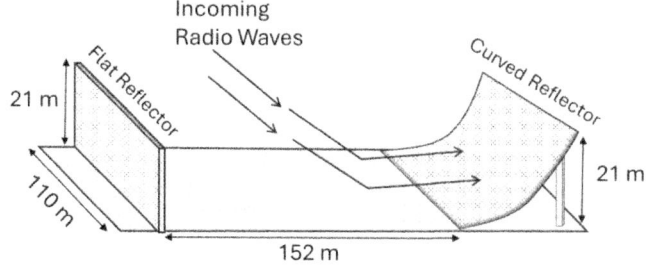

Figure 10. Schematic of the Big Ear Radio Telescope. You can find nice real images here: https://www.bigear.org/slideshow/tour.htm

The telescope was automated, churning out endless reams of computer printouts filled with columns of numbers and letters. On the 15th of August 1977, that patient, immobile machine caught something extraordinary; however, there was no one there to witness as it happened. Instead, days later, in a cramped office stacked with paper, volunteer astronomer Jerry Ehman sat down to do what seemed like a boring task: comb through the data line by line, hunting for anomalies. He had done this countless times before, usually finding nothing but noise.

Then, his eyes froze on a peculiar string of characters: **6EQUJ5**

The numbers and letters represented signal strength over time. To Ehman's trained eye, this wasn't random static — it was a burst of radio energy far stronger than anything expected from natural background noise. Without thinking, Ehman grabbed his red pen, circled the sequence, and scribbled a single word in the margin:

"Wow!"

That impulsive annotation gave the signal its immortal name.

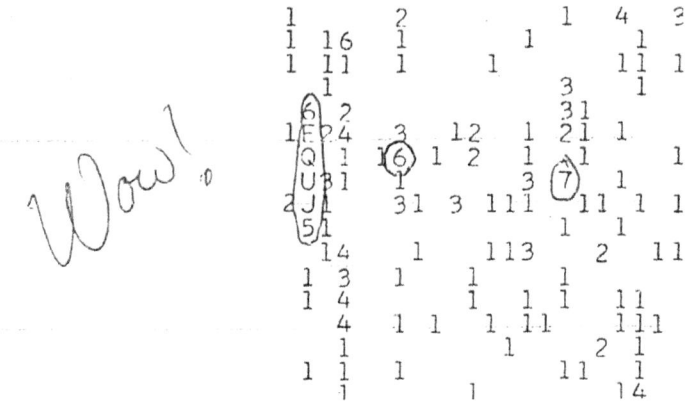

Figure 11. The Wow! Signal (File:Wow signal.jpg - Wikimedia Commons)

To this day, the Wow! Signal remains one of SETI's most tantalizing mysteries. Was it a natural phenomenon, or a deliberate whisper from another mind?

Imagine you're trying to find out if a civilization has radio technology. You don't need to send a long message. You don't need to say "hello." All you need is a single burst, right where they're most likely to be paying attention. If they notice, you've learned something important: *they are listening*.

If we assume the Wow signal was deliberate, the implication is that this hypothetical civilization would already know our approximate location. It is not far-fetched to think someone would know we exist, considering that we had been unintentionally sending radio waves to outer space since the early 20th century. Anyone within a radius of about 77 light years could have detected those signals.

Could the Wow! signal have been a test? A probe, not of our planet, but of our awareness. Would we hear? Would we recognize it? Would we respond?

We did hear. We did recognize. But we did not respond. However, even without a response, our awareness could have been detected by observing changes in our radio emissions, media coverage, etc. Our silence would have spoken volumes. Whoever sent the burst — if it was sent — now would know two things:

1. Earth has radio astronomy capabilities.
2. Earth is cautious, perhaps unwilling to answer.

It is important to note that the lack of repetition of the signal makes it hard to prove intent. A true monitoring plan would likely send multiple bursts to confirm detection. However, perhaps our silence may have shaped what came next. Because years after, the cycle didn't stop. It shifted from radio to physical visitors. If the first test was sound, the next one would be sight. A civilization probing our awareness wouldn't stop at one channel. They would escalate — from a whisper in the radio band to a messenger crossing our skies.

Stage 2, Surveillance: are we observing? 1I/Oumuamua

Forty years after the Wow! signal, the sky delivered something new. In October 2017, astronomers in Hawaii spotted a faint, fast-moving speck that didn't belong to the solar system. It was named 1I/'Oumuamua, the 1I describes it as the first interstellar visitor, and the name is a Hawaiian word meaning "a messenger from afar arriving first."

This object was unlike anything we had seen:

- **Shape:** Extremely elongated, perhaps ten times longer than it was wide — described as a cosmic cigar tumbling end over end.
- **Behavior:** It showed a slight non-gravitational acceleration after passing the Sun, but without the gas jets of a comet. Some joked it was a "light sail," others a "cosmic prank cigar."
- **Brightness:** It showed extreme brightness variations (up to a factor of 10) which made astronomers joke it was "twirling like a baton."
- **Timing:** It arrived almost exactly 40 years after the Wow! signal — one human generation later.
- **Location:** The Wow! signal appeared to come from Sagittarius, near the galactic center, while 1I/'Oumuamua arrived from the direction of Lyra/Vega.

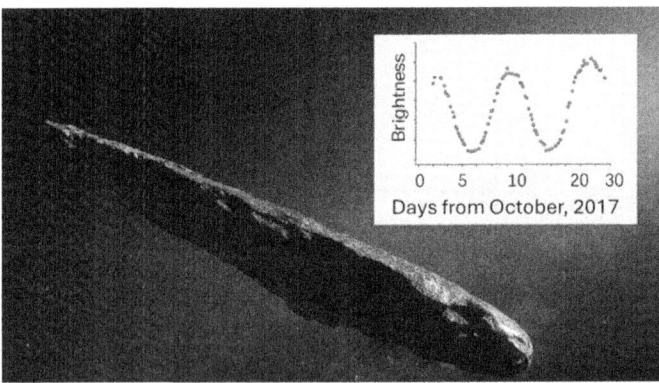

Figure 12. 1I/Oumuamua. Artist impression, showing extreme brightness variations.

That generational cadence is striking: If one imagines a designer scenario, 40 years could be interpreted as a deliberate cadence — long enough to observe how humanity evolved after the Wow! signal, yet short enough to feel intentional. But this timing could just as easily be coincidence. In 1977, radio SETI was still young, and the Wow! signal became its most famous anomaly. 40 years later, in 2017, astronomy had advanced dramatically — wide-field surveys like Pan-STARRS were capable of spotting fast-moving, faint interstellar objects. The timing almost feels like a "next stage" challenge. The differing arrival directions are intriguing. If one entertains a speculative scenario, they could be interpreted as part of a distributed observational pattern — not a repeat from the same source, but a new arrival from another region of the sky. Of course, this could also be simple celestial happenstance.

Could we spot a fleeting visitor from beyond the stars? Could we recognize its strangeness? We did. Telescopes worldwide pivoted, papers poured out, and debates erupted. The messenger had been seen.

It is important to point out that, if 1I/'Oumuamua were engineered, its trajectory was not optimized for close study of Earth — it passed quickly and was faint. A true probe would likely linger or maneuver. The most intriguing part of it was its non-gravitational acceleration as it passed close to the sun, without any detectable outgassing, this still defies an explanation.

Stage 3, Baseline sampling: the ordinary stranger 2I/Borisov

Just two years after Oumuamua, in August 2019, another interstellar traveler appeared. This time it looked familiar: a comet with a glowing coma and a tail stretching millions of kilometers. Discovered by amateur astronomer Gennadiy Borisov in Crimea, it was named 2I/Borisov — the second interstellar object ever found.

Unlike 1I/'Oumuamua 's bizarre shape and non-gravitational acceleration, 2I/Borisov looked and behaved like a textbook comet: a nucleus, coma, and tail. It was the first confirmed interstellar comet, showing that not all visitors are anomalies. Nevertheless, its chemistry was unusual: astronomers found it rich in carbon monoxide, suggesting it formed in an extremely cold environment, perhaps around a red dwarf star.

2I/Borisov arrived just two years after 1I/'Oumuamua . That short gap feels almost like a deliberate follow-up: first an anomaly, then a normal case, as if to test whether scientists would overreact or stay cautious. In our own scientific discovery, we often perform "control tests", which serve one purpose: to provide a baseline for comparison. These are essential for establishing validity, reliability, and cause-and-effect relationships.

In a laboratory, when an experiment produces a surprising result, scientists immediately seek a control — a baseline case that shows what "normal" looks like under similar conditions. 2I/Borisov could have been that baseline. Proving that interstellar objects can be ordinary comets, formed in distant star systems and wandering into ours. By providing a control, 2I/Borisov sharpened the contrast, one could imagine —

purely speculatively — that a normal case might appear after an anomaly to keep observers cautious so that anomalies like 1I/'Oumuamua and 3I/ATLAS stand out more sharply. But, of course, a far more likely explanation is that 2I/Borisov was simply natural, and our reaction to it was the only thing worth monitoring.

Could we tell the difference between natural and anomalous? Would we dismiss the first messenger as a fluke, or recognize that even "ordinary" visitors carried extraordinary implications? 2I/Borisov could have served as proof that interstellar comets exist, and that the cosmos delivers both anomalies and norms. From a monitoring perspective, it's unlikely an advanced civilization would send or deflect a natural-looking comet just as a test. More likely, 2I/Borisov is simply natural, but our reaction and analysis of it could still have been monitored in the same ways as described before.

Figure 13. 2I/Borisov, artistic representation

Stage 4, Pattern analysis: the real stranger 3I/ATLAS

By 2025, astronomers were no longer surprised by the idea of interstellar visitors. The next one was 3I/ATLAS, discovered in 2025. It was discovered by the Asteroid Terrestrial-impact Last Alert System (3I/ATLAS) in Hawaii. At first glance, it seemed ordinary — another comet from beyond the solar system. Yet its timing and character carried a deeper resonance.

According to calculations from Avi Loeb, Its radiant lay just nine degrees from the Wow! signal's pointing in Sagittarius — only about 18 Moon widths apart. Unlike 'Oumuamua, 3I/ATLAS hugged the ecliptic, the plane of planetary orbits, as if deliberately tracing the same band of sky where the Wow! signal had once whispered. 3I/ATLAS was active, with a coma and tail. It seemed to tie the cycle together: radio first, then cometary probes, all along the same corridor.

Compared to 2I/Borisov, 3I/ATLAS is a real enigma. It breaks almost everything we know about comets (interstellar or not). There are at least 14 unexplained behaviors that make 3I/ATLAS stand out from the crowd. Each of these anomalies can be explained by natural processes but together really make scientists scratch their heads trying to avoid including a more exotic explanation in their models — the possibility that 3I/ATLAS could be artificial. This remains speculative, but the anomaly stack makes the question difficult to ignore:

Would we notice that its path echoed the signal from 1977? Could we differentiate a textbook comet like 2I/Borisov from a probe concealed as a natural object?

Figure 14. 3I/ATLAS, artistic representation

The Jupiter blind test

Up to this point, we've been looking backward — analyzing the anomaly stack, weighing competing explanations, and considering what 3I/ATLAS *has already done*. But the story is not finished. 3I/ATLAS is still moving, still evolving, and its next destination offers something rare in science: a natural experiment we did not design but can still learn from.

3I/ATLAS is now on course toward Jupiter, the largest gravitational well in the Solar System. That alone is remarkable. But what makes this moment scientifically precious is that Jupiter acts as a kind of cosmic amplifier. Its gravity, radiation belts, magnetic field, and plasma environment stress-test every object that passes near it. Natural comets behave one way under that stress. Controlled objects behave another. And objects with intent — whether reconnaissance or something more — behave differently still.

In other words, Jupiter gives us a **blind test** (an experiment where participants are kept unaware of key information to avoid bias). A chance to watch what 3I/ATLAS does next, without knowing the answer in advance.

To make this test meaningful, we can outline what each of the four scenarios would predict — not with certainty, but with enough structure that future observations can shift the Bayesian balance shown in the first chapter. What would we expect for each scenario?

Scenario 1: Preparatory test for first contact

A sequence of escalating probes designed to measure our awareness, interpretive discipline, and technological maturity before any direct communication. What this scenario would predict near Jupiter:

1. Purposeful but subtle maneuvering

- Small course corrections that keep 3I/ATLAS intact
- Trajectory shaping that looks "optimized" rather than random.
- Avoidance of high gravitational stress regions

2. Behavior that seems designed to be *interpretable*

- Orientation changes that look like deliberate pointing
- Activity patterns that are structured rather than chaotic

- Brightness variations that repeat or stabilize

3. A trajectory that maximizes scientific observability

- Passing through regions where Earth-based instruments can best track it.
- Alignments that seem "too convenient" for data collection

4. No overt signaling — but no attempt to hide

This scenario assumes they want us to *notice patterns*, not panic.

In short: If 3I/ATLAS behaves like something running a controlled experiment on our interpretive abilities — subtle, structured, and non-random — this scenario gains strength.

Scenario 2: Passive Observation

A long-term, non-intrusive watcher that studies us the way we study ecosystems — quietly, without interference.

What this scenario would predict near Jupiter:

1. A trajectory that prioritizes safety over efficiency

- Smooth, low-thrust corrections.
- Avoidance of high-risk gravitational zones
- No dramatic maneuvers

2. Stability under stress

- No fragmentation
- No chaotic jet behavior

- No sudden changes in rotation

3. Orientation changes consistent with scanning

- Slow, deliberate reorientation
- Brightness changes that match sensor-pointing behavior
- No thrust signatures that look like propulsion

4. A post-Jupiter path that continues a long observational arc

- Outbound trajectory toward the outer Solar System
- No approach toward Earth or inner planets

In short: If 3I/ATLAS behaves like something trying to *stay intact and keep watching*, this scenario strengthens.

Scenario 3: A Preamble to a Threat

A reconnaissance pattern optimized not for communication or curiosity, but for mapping vulnerabilities, blind spots, and response behaviors.

What this scenario would predict near Jupiter:

1. A trajectory that exploits Jupiter for strategic advantage

- A gravity-assist maneuver that increases speed or changes direction sharply.
- A path that looks optimized for coverage, not safety

2. Jet behavior that resembles controlled propulsion

- Stable, collimated jets
- Acceleration that increases near closest approach
- Course changes inconsistent with natural outgassing

3. Behavior that tests our detection and response capabilities

- Maneuvers that occur during known observation windows
- Sudden changes that appear to probe our tracking systems
- Activity spikes that look like "pings" (test signals) rather than natural outbursts

4. A post-Jupiter trajectory that approaches strategic regions

Examples:

- Earth's orbital plane
- The inner Solar System
- The heliosphere boundary
- Orbits that allow repeated passes

In short: If 3I/ATLAS behaves like something mapping our environment or testing our reactions, this scenario becomes more plausible.

Scenario 4: A Purely Natural Phenomenon

A strange but ultimately natural interstellar object. What this scenario would predict near Jupiter:

1. A purely gravitational trajectory

- No course corrections
- No thrust-like acceleration
- Path matches celestial mechanics models exactly.

2. Increased activity due to tidal or thermal stress

- Jets intensify unpredictably.
- Dust production increases.
- No directional or structured behavior

3. Possible fragmentation

Natural comets often break apart near Jupiter.
If 3I/ATLAS fractures chaotically, that strongly supports the natural model.

4. No purposeful orientation changes

- Spin state remains noisy.
- Brightness variations remain stochastic.
- No signs of pointing or scanning.

In short: If 3I/ATLAS behaves like a fragile, volatile-rich body being bullied by Jupiter's gravity, the natural scenario gains strength.

Table 3. Summary of expected behaviors for the Jupiter Blind Test

Scenario	If this scenario, 3I/ATLAS might...
1. Preparatory Test for First Contact	• Make small, smooth course adjustments that look intentional rather than random • Change its orientation in ways that seem "purposeful," like it's pointing or scanning • Follow a path that makes it easier for us to observe • Show patterns in brightness or activity that look structured, not chaotic
2. Passive Observation	• Avoid dangerous areas and take the "safe route" around Jupiter • Stay stable and intact with no dramatic behavior • Slowly reorient itself as if it's watching or measuring something • Continue on a calm, steady path out of the Solar System afterward
3. Preamble to a Threat	• Use Jupiter to make a sharp turn or speed boost, like a deliberate maneuver • Show jet activity that looks like controlled propulsion, not random outgassing • Make sudden changes that seem to test our tracking abilities • Head toward a strategically important region afterward (e.g., inner Solar System)
4. Purely Natural Phenomenon	• Follow the exact path predicted by gravity with no surprises • Become more active or unstable as Jupiter's gravity stresses it • Possibly break apart or shed material in a messy, chaotic way • Show no signs of control, structure, or purposeful behavior

The Caravan ahead

If 3I/ATLAS is natural, then this chapter ends as a scientific curiosity — a reminder that the universe still surprises us. But if it is not natural, then perhaps it is not alone. Perhaps 3I/ATLAS is only one member of a longer procession, a caravan of visitors spaced across decades, each one offering a different kind of puzzle: a radio whisper, a silent scout, a baseline comet, and now an object whose behavior forces us to confront our own interpretive limits.

If one imagines this as a design — and that remains speculative — then perhaps the sequence is not finished. If this sequence reflects something intentional, the next arrival could already be inbound, hidden among the faint specks drifting toward the Sun. It might look ordinary. It might look extraordinary. It might be disguised so well that only patience, pattern recognition, and intellectual discipline will reveal its purpose.

The deeper question is not whether another visitor will come. It is whether we will notice. Will we connect the dots, or treat each anomaly as an isolated curiosity? Will we recognize choreography, or insist on coincidence? Will we see the pattern only in hindsight when the caravan has already passed? The Wow! signal may have been the first knock on our door, 1I/'Oumuamua may have been the second, 2I/Borisov the third, 3I/ATLAS the fourth.

As 3I/ATLAS approaches Jupiter, we are about to witness a rare blind test. Its behavior in the coming months may sharpen the picture: whether this is a natural wanderer, a passive observer, a reconnaissance probe, or the early stages of something more deliberate.

The next stage may already be on its way, timed to arrive when we least expect it. **Could the next step — if there is one — be something closer to contact, or simply the next clue in a sequence we are only beginning to interpret?**

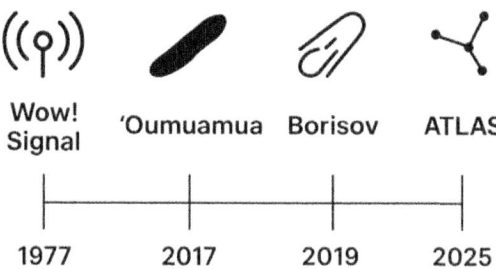

Figure 15 Sequence of events. So far...

Chapter 3: Are we being tested?

A test is fine... but we never got the notes.

By now, the anomalies no longer feel like isolated curiosities. The anomaly stack from the previous chapter shows us something unsettling: these visitors do not just differ in their physics. They differ in their *roles*. One tests our ability to hear. Another tests our ability to track. Another tests our ability to distinguish natural from artificial. And the latest tests our ability to coordinate across blind spots and uncertainty. It is as if each arrival probes a different dimension of our awareness.

This chapter asks a question that now feels unavoidable: What if these events are not random? What if they are part of a deliberate progression — a stepwise test of our observational, psychological, and collective maturity?

This is not a claim. It is a thought experiment. But it is a thought experiment worth taking seriously because the pattern we see mirrors the very reconnaissance logic we use when approaching the unknown. Before contact comes calibration. Before dialogue comes assessment. Before trust comes testing.

If someone — or something — is evaluating us, then the sequence of anomalies may not be a coincidence at all. It may be a **protocol**. And if that is true, then the real subject of the test is not the objects in the sky.

What kind of test?

If this is a test, what's the subject? Could it be:

- **Surveillance:** Cataloguing which civilizations notice and which don't.

- **Science:** Studying how intelligent species react to anomalies.

- **Reconnaissance:** Mapping our defenses, our awareness, our blind spots.

- **Contact rehearsal:** Seeing if we respond, and how.

Tests, if that's what these are, imply the possibility of a tester. If these visitors are part of a cycle, then someone — or something — is watching. Not constantly, not loudly, but quietly, through probes that look like comets and signals that vanish after seconds.

If one imagines a surveillance scenario, it would be subtle rather than overt —not invasion, not contact, just observation. One could imagine — speculatively — a scenario resembling a galactic census. The reader would wonder why an active invasion scenario is not included in the above list. But if we follow intent as a logical argument. Why would anyone travel such vast distances and bother with all these tests if their intention was to invade? Natural resources such as minerals and water are abundant in the universe. Life is the only thing that could set Earth apart from most planets. Any intelligence capable of such feats would also have the technology to annihilate us—so why bother with tests?

If this were a test, the natural next question would be: **what happens if we 'pass' — or fail?** Do we get ignored, approached, or judged? The cycle suggests that the next visitor will not just test our awareness, but our **wisdom**. And that is where the mystery becomes personal. Because the test is no longer about telescopes or radio dishes. It's about us.

Stepwise reconnaissance protocol

When we ask whether humanity is being tested, it helps us to recognize that **stepwise reconnaissance** is common to us across science, medicine, and military practice, humans have long relied on staged protocols to probe the unknown. Each field uses a similar logic: begin with minimal risk, escalate only when necessary, and build layer by layer toward deeper understanding.

In science, reconnaissance begins with broad surveys. Planetary missions, for example, rarely start with landers. They begin with flybys to map the terrain, then orbiters to refine measurements, and only later commit to rovers or sample-return missions. This stepwise approach ensures that anomalies are noticed, baselines are established, and risks are managed before deeper engagement. It is a ladder of awareness: survey → probe → control → escalation.

In medicine, the same principle governs diagnosis. Physicians start with non-invasive tests — history, physical exam, basic labs — before escalating to imaging, biopsies, or invasive monitoring. Each stage is designed to minimize harm while maximizing insight. Critical care protocols are explicitly stepwise: escalate monitoring only if the patient's condition demands it. Medicine treats the body as an unknown environment to be explored cautiously, probing deeper only when the evidence justifies it.

In the military, stepwise reconnaissance is a doctrine. Armies begin with intelligence and aerial surveillance to detect presence. They escalate to patrols or flybys to test detection and tracking. Next comes baseline observation — watching routine activity to distinguish normal from anomalous. Finally, they deploy close probes or contact reconnaissance, sometimes disguised or fragile, to stress test coordination and resilience. Each stage is a calculated escalation, designed to reveal awareness, blind spots, and response capacity without committing full force.

Altogether, these contexts show a unifying thread: **structured escalation under uncertainty**. Whether probing a planet, diagnosing a patient, or mapping an adversary, humans rely on stepwise reconnaissance to minimize risk while maximizing information. Each stage is a test — of awareness, of skepticism, of resilience, of coordination. As shown in the list below:

Stage 1: Informational probe (ping the environment)

- **When we'd do it:** Before committing resources, we'd send a simple, universal signal — something any advanced species could recognize (radio pulses, laser flashes, neutrino bursts).

- **Purpose:** Test if they're listening. This is the cheapest, lowest-risk way to gauge awareness.

- **Analogy:** Like sonar pings before entering unknown waters — "Are they listening?"

Stage 2: Material Reconnaissance (Flyby probe)

- **When we'd do it:** Once we know they can hear, we'd send a fast, hard-to-miss probe past their skies.
- **Purpose:** Test their ability to track and characterize anomalies.
- **Analogy:** Military reconnaissance aircraft buzzing a border — "Are they watching? Can they follow motion and measure anomalies?"

Stage 3: Control Sample (Baseline object)

- **When we'd do it:** To avoid confusion, we'd send a "normal" object— or natural-looking artifact.
- **Purpose:** Provide a calibration sample so they can distinguish natural vs engineered.
- **Analogy:** In science, we always include a control group. In reconnaissance, we'd test whether they can separate noise from signals.

Stage 4: Deployment Probe (Complex test object)

- **When we'd do it:** After confirming awareness and skepticism, we'd escalate. Send a fragile, complex probe designed to fragment, disguise itself, or reactivate later.
- **Purpose:** Test resilience, coordination, and long-range tracking.
- **Analogy:** A stealth drone that deliberately malfunctions, forcing the target to coordinate across multiple sensors to reconstruct what happened.

Why this protocol is used.

- **Risk management:** Each stage escalates slowly, minimizing exposure while maximizing intelligence.

- **Capability mapping:** We'd learn if the other civilization has ears (radio), eyes (tracking), skepticism (distinguishing anomalies), and coordination (multi-platform surveillance).

- **Psychological test:** By timing concealment and reactivation, we'd probe not just technology but decision-making under uncertainty.

- **Strategic outcome:** After four stages, we'd know whether they're cautious, curious, coordinated, or blind — without ever revealing our full hand.

When we look at the Wow! signal, 1I/'Oumuamua, 2I/Borisov, and 3I/ATLAS through this lens, the resemblance is evident. The sequence mirrors our own reconnaissance logic: a simple ping to test if we listen, a fast probe to test if we watch, a control sample to calibrate our skepticism, and a fragile, complex body to test our coordination. If these events are deliberate, they follow the same protocol we ourselves would design before approaching another civilization. Which raises the unsettling possibility that this pattern could resemble a test (speculatively speaking).

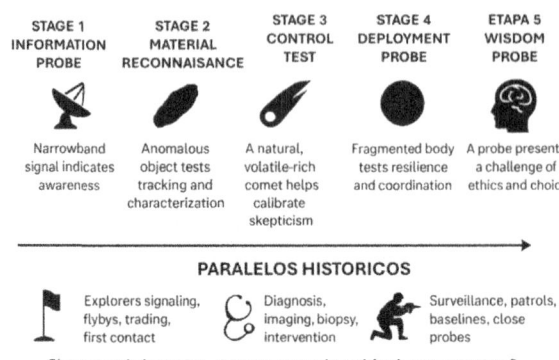

Figure 16. Stepwise reconnaissance protocol

Figure 17 Would a tester follow the same strategies as we would use to test others?

Historical parallels

Throughout history, powerful civilizations have tested others in **stepwise ways** before deeper engagement:

- **Age of Exploration (15th–17th centuries):** European ships often lingered offshore before landing, sending signals or emissaries to gauge awareness. The first encounters were informational probes —flags, bells, or cannon shots— testing whether coastal societies noticed.

- **Colonial Contact:** Traders introduced "control samples" of goods (beads, textiles, iron tools) to see how locals reacted before escalating to settlement. These were baselines against which anomalies (guns, ships, diseases) stood out.

- **Cold War Reconnaissance:** Superpowers used staged surveillance — satellites to test "ears," spy planes to test "eyes," and controlled incidents (like submarine surfacing) to probe coordination. Each step escalated awareness without triggering outright conflict.

These parallels show that **stepwise reconnaissance is a human instinct**: we probe cautiously, escalate deliberately, and measure responses before committing.

Figure 18 Common uses of stepwise reconnaissance protocols.

Philosophical reflections

The unsettling symmetry is that the **tests we suspect are the very tests we ourselves would design.** This raises deeper questions:

- **Tests as Mirrors:** If we are being tested, the protocol reflects our own methods of probing the unknown. The universe may be holding up a mirror, showing us how predictable our logic is.

- **Awareness vs. Wisdom:** Awareness is easy to measure — telescopes, radio dishes, orbital tracking. Wisdom is harder. Do we respond with curiosity or fear? Do we interpret anomalies with open minds or dismiss them as noise?

- **The Ethics of Testing:** In medicine, reconnaissance is justified by care. In the military, it is justified by survival. In science, it is justified by curiosity. But if another intelligence is testing us, what is their justification? Cataloguing? Surveillance? A galactic census? Or something more evaluative?

Philosophically, the tests force us to confront not just our instruments, but our **values**. What does it mean to "pass" or "fail" a cosmic test?

If — hypothetically — this were a linked series of events designed to test our reactions, this is how that scenario could be summarized:

Stage 1: Informational Probe (1977 Wow! Signal)

- **Event:** Narrowband hydrogen-line emission detected at Ohio State's Big Ear telescope. This was a simple, elegant signal—brief, powerful, and precisely tuned to a frequency that any advanced civilization would recognize.

- **Function (in a speculative scenario):** Test whether Earth "is listening" — can we detect and recognize a deliberately simple beacon?

- **Outcome:** We noticed but never replied. The signal became famous, yet remained a mystery.

- **Implication:** If it was deliberate, the senders now know we have radio astronomy, but also that we are cautious or silent.

Stage 2: Material Reconnaissance (2017, 1I/'Oumuamua)

- **Event:** The first interstellar object detected— elongated, shard-like morphology, and unexplained accelerations. Its brightness varied dramatically, suggesting a tumbling motion. Its trajectory was hyperbolic, confirming it came from beyond the solar system.

- **Function:** Test whether Earth is "watching"—can we track and characterize a fast, anomalous body?

- **Outcome:** We noticed and scrambled telescopes worldwide, but data was limited. Morphology and motion remain unexplained.

- **Implication:** Senders now know we can detect interstellar artifacts, but our observational cadence leaves gaps.

Stage 3: Control Sample (2019 2I/Borisov)

- **Event:** The second interstellar object behaved like a textbook comet—volatile-rich, predictable, and familiar

- **Function (if interpreted through this lens):** Provide a "baseline" natural visitor to contrast with engineered anomalies. A test designed to show what normal looks like, so the next visitor's anomalies stand out.

- **Outcome:** We studied it thoroughly, confirming our ability to characterize natural interstellar comets.

- **Implication:** Senders now know we can distinguish "normal" from "anomalous" — a calibration test.

Stage 4: Deployment Probe (2025 3I/ATLAS)

- **Event:** Showed odd chemistry and jets, non-gravitational motion. Its trajectory could be interpreted as unusually well-aligned —as if calculated— with the planetary orbital plane and Jupiter's gravitational field.

- **Function:** Test whether Earth can study a fragile, natural-looking body that may conceal a deeper structure by passing perihelion at the worst possible time for us to observe.

- **Outcome:** We noticed and recovered partial data. We captured imagery, spectroscopy, and refined trajectory. We continue monitoring and observing closely with all available technology.

- **Implication:** Senders would know we can coordinate multi-platform surveillance across planets — but that our coverage still has blind spots. Path toward Jupiter may be a rendezvous or tracking test; releasing smaller probes near Earth or other planets remains a possibility.

Speculative future stages

If the cycle continues, Stage 5 and beyond may probe dimensions beyond awareness:

Stage 5: Wisdom Probe

- **Event:** A visitor that encodes cultural or ethical data — perhaps a signal carrying stories, mathematics, or moral dilemmas.

- **Function:** Test whether Earth can interpret not just anomalies but meaning.
- **Outcome:** Our response would reveal whether we are ready for dialogue, not just detection.
- **Implication:** Passing means demonstrating intellectual honesty and interpretive skill; failing means being catalogued as unready.

Stage 6: Biosphere Probe

- **Event:** A probe that interacts with Earth's atmosphere or biology in subtle ways — perhaps seeding microbes or testing planetary resilience.
- **Function:** Measure how we monitor and protect our biosphere.
- **Implication:** A test of stewardship, not just awareness.

Stage 7: Contact Rehearsal

- **Event:** A signal or artifact that forces a choice: reply or remain silent.
- **Function:** Test our collective decision-making under pressure.
- **Implication:** Reveals whether humanity can act as one civilization, or fracture under stress.

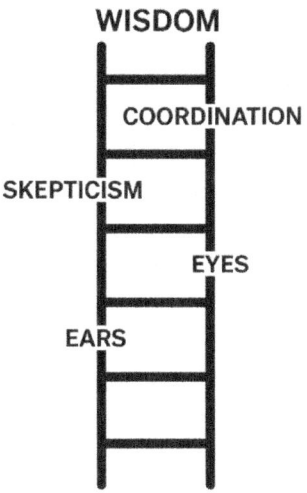

Figure 19. Is the final goal to assess our wisdom?

Why would they want to test us?

The biggest question of all is why? What purpose would it serve to send signals, shards, comets, and spheres across the void, just to test our awareness? Motives are in the shadows.

There are several possibilities, each unsettling in its own way:

- **Surveillance:** Imagine a galactic census. Civilizations are catalogued, their awareness tested, their reactions noted. Not to invade, not to contact, but simply to record: *this species hears, sees, compares, reacts.*

- **Science:** Perhaps we are part of an experiment. Just as we study animals in the wild, maybe someone is studying us — observing how we respond to anomalies, how we publish data, how we argue among ourselves.

- **Reconnaissance:** Another possibility is strategic. If probes are left behind, they may be mapping our resources, our defenses, our blind spots. Not to strike, but to know. Knowledge is power, even across the stars.

- **Contact rehearsal:** Or maybe this is practice. A rehearsal for eventual communication. First a signal, then a shard, then a control, then a deployment. Each step closer to seeing how we handle the unknown.

What makes these speculative motives intriguing is that they mirror our own behavior. When we send probes to Mars or Europa, we don't announce ourselves to microbes. We observe quietly, measure carefully, and test environments before deciding what to do next. If we act this way toward smaller life, why wouldn't a more advanced civilization act this way toward us?

Notice what's missing: no messages, no greetings, no declarations. Just silence. Silence is safer. Silence lets the watchers learn without revealing themselves. And silence forces us to wrestle with the mystery on our own.

If we truly are being tested, then the next question is no longer about them at all—it's about us. Because tests reveal the subject, not the examiner. If an intelligence is watching, then every anomaly becomes a mirror held up to our species. Our fear, our curiosity, our skepticism, our coordination, our aggression — these are the variables being measured. The silence we find so unnerving may not be indifference, but evaluation. And if that is the case, then the real mystery is not what the watchers are doing, **but what they are learning**. To understand the test, **we must understand what we are showing them.**

That is where we turn next.

Chapter 4: What would they observe?

I just wish we'd had time to tidy up first.

Let's take a step back for a moment. Picture Earth from far away — a blue dot orbiting a yellow star, tucked into the spiral arm of the Milky Way. To us, it feels vast and full of mystery. But to someone watching from afar, it might look small, fragile, and fleeting. And if the cycle of visitors was real then we would already be part of something larger. We might be noticed, catalogued, tested. Not attacked. Not contacted. But observed. That alone changes how we see ourselves. To be catalogued means we are no longer invisible. Our civilization would have crossed a threshold: we are loud enough, bright enough, and aware enough to be worth recording. That's not a threat. It's a milestone. It means we've joined the list of species that matter — at least to someone, somewhere. Because if the watchers were real, then the next phase of the test wouldn't be about astronomy, but about psychology.

With visibility comes responsibility. If we are being watched, then our behavior is part of the data. Our reactions, our coordination, our maturity — all of it becomes part of the record. If this is the case, then the real question is not just what we see in the sky, but how we act on Earth.

History shows us that when humans realize they are being observed — whether by nature, by other civilizations, or by one another — our responses fall into recognizable patterns. These patterns reveal not only our fears and hopes, but also our maturity as a species.

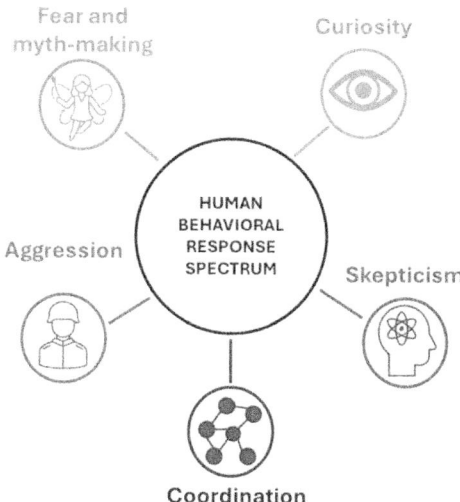

Figure 20. An external observer would see not just our technologies, but our behavioral instincts. Fear, curiosity, skepticism, coordination, and aggression form the core axes of human response to the unknown. Each instinct carries both strengths and vulnerabilities, and together they create the psychological profile we project into the cosmos.

In this context, every anomaly becomes a kind of psychological mirror — a way of revealing who we are under pressure. Fear, curiosity, skepticism, coordination, aggression: these are not random reactions. They are the core behavioral instincts of our species, the patterns that would stand out most clearly to any intelligence observing us from afar. And just as biologists classify animal behavior into measurable traits, we can do the same for ourselves.

To understand what an external observer would learn from our responses, we need a simple framework — a set of axes that capture the strengths, vulnerabilities, and signals we project into the cosmos. What follows is not a judgment, but a map: a way to see ourselves as others might see us.

Instinct 1: Fear and myth-making

The first human instinct is often fear. Unknown lights in the sky have long been interpreted as omens. Ancient civilizations saw comets as harbingers of disaster, eclipses as signs of divine anger, and sudden anomalies as portents of war. Fear is a natural reaction to uncertainty: it protects us by preparing for danger. But fear also clouds judgment. It can turn a neutral event into a perceived threat.

Modern society is not immune to myth creation. Rumors spread quickly, amplified by media and social networks. A strange signal or object can trigger speculation ranging from alien invasion to government conspiracy. The watchers — if they exist — would note this tendency. They would see that humanity often interprets anomalies through the lens of fear before curiosity has time to take hold.

In the modern era, fear does not spread through omens or village rumors — it spreads through social media. Social media can be both a blessing and a curse for science, from one side, it gives scientists a new means for expanding their ideas (access of information) and communicating to the public (e.g., Niel de Grass Tyson's Star Talk channel has nearly 3.6 million followers on Facebook alone). However, social media platforms are designed to maximize engagement, reward speed, emotion, and simplicity. A dramatic headline, a blurry photo, or a speculative claim can travel across the globe in minutes. By the time scientists publish careful analyses, the non-scientific narrative has already hardened the public imagination. This creates a dangerous asymmetry:

- **Speed vs. Accuracy**: Misinformation travels faster because it is simple, emotional, and shareable. Scientific analysis travels slower because they require data, peer review, and cautious language.

- **Emotion vs. Evidence**: Fear-based posts trigger strong reactions — outrage, anxiety, excitement — which drive clicks and shares. Evidence-based posts often use technical language, charts, or caveats, which feel less compelling to a general audience.

- **Virality vs. Verification:** Algorithms amplify content that spreads quickly, regardless of truth. Verification takes time, and corrections rarely reach the same audience as the original claim.

Scientific anomalies, whether ordinary or extraordinary, are rapidly transformed into cultural myths because sensational narratives spread far faster than careful explanations. When 1I/'Oumuamua appeared, its unresolved anomalies created a perfect vacuum for speculation. When this happens, scientists are quick to urge restraint and offer cautious and incomplete natural explanations, but this very caution frustrates many outside the field. To non-experts, the absence of definitive answers feels evasive, and turns scientific uncertainty into perceived avoidance and accelerates mistrust. As a result, viral narratives fill the gap long before peer-reviewed papers can catch up.

By the time 3I/ATLAS arrived, the ecosystem of speculation was fully primed and the same situation repeated. Scientists produced only a handful of papers, while millions of posts, videos, and memes shaped the public narrative. Scientific discourse is cautious, incremental, and often opaque, while public narratives thrive on immediacy, certainty, and drama. The result is a widening gap in which scientific restraint is perceived as dismissal, and the slow pace of evidence-based communication inadvertently fuels the very myths it seeks to correct.

Why this matters

This dynamic explains why watchers —if they were real— might test us first instead of contacting us directly. A direct message risks triggering panic, confusion, or myth-making that spirals beyond control. Ambiguous anomalies, however, let them observe our reactions safely. Do we panic? Do we invent legends? Do we fracture into mistrust? Or can we balance skepticism with imagination? By watching how we respond to uncertainty, they could assess whether humanity is capable of maturity before risking dialogue.

In a darker interpretation, potential threats could exploit this same vulnerability. Fear and myth-making can function as tools of psychological warfare. By releasing anomalies that look threatening but remain unexplained, they could provoke paranoia, conspiracy, and institutional mistrust. Governments might be accused of hiding the truth. The trust in the scientific community would be lost. Public discourse could splinter into competing myths. No weapon would be needed; destabilization would come from within, driven by our own narratives. In such a scenario, fear becomes the battlefield, and our imagination becomes the weapon turned against us.

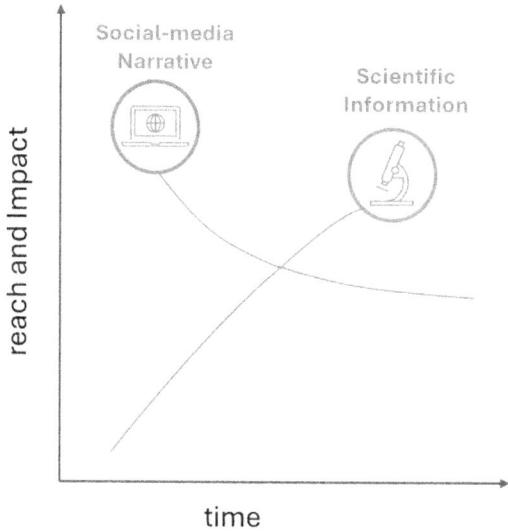

Figure 21. This figure contrasts the speed and reach of social-media narratives versus scientific analysis. The orange curve shows how emotional, myth-driven content spikes rapidly, often dominating public discourse before scientific explanations emerge. The blue curve shows the slower, more cautious rise of evidence-based analysis. Eventually, both narratives merge into one.

Instinct 2: Curiosity and exploration

The second instinct is curiosity. Human beings are explorers by nature. Once fear subsides and skepticism finds balance, the impulse to look beyond takes over: to measure, compare, classify, and explain. This inclination is not a modern luxury but a force that has shaped our history from the very beginning.

Civilizations separated by oceans and millennia independently developed advanced mathematics, astronomical calendars, and models of the sky. The **Sumerians** invented numerical systems and recorded planetary motions with astonishing precision. The **Egyptians** aligned temples and pyramids with stars and solstices, blending architecture and cosmology into a single act of observation. The **Maya** built extraordinarily accurate calendars, predicting eclipses without telescopes or modern instruments. In **China**, imperial astronomers catalogued comets, Novas, and sunspots for centuries. In **India**, mathematicians developed trigonometric techniques that later influenced the entire world.

None of these cultures knew of each other, yet all shared the same impulse: **to look upward, search for patterns, and try to understand them**. Curiosity is a universal language of our species.

This trait has been a decisive advantage. It allowed us to navigate oceans, harness fire, invent writing, decipher the cycles of the seasons, and eventually build telescopes, particle accelerators, and interplanetary probes. Curiosity is the root of science, but also of technology, art, and philosophy. It is the force that drives us to ask "why?" and "what lies beyond?"

But curiosity also has a more complex side. At times it has led to conflict, rivalry, and tension. The pursuit of knowledge has fueled explorations that ended in conquest; it has produced discoveries later used as weapons; it has sparked religious, political, and cultural disputes when new ideas collided with established beliefs. Curiosity opens doors, but it can also destabilize what we take for granted.

Even so, it remains one of our most defining traits. We are a species that cannot help but look, measure, investigate. Curiosity is our oldest compass — a source of discovery, and at times, of risk.

Why this matters

Curiosity is one of our greatest strengths — but in a scenario of evaluation, it could also become an open door. A hypothetical observer could exploit that trait with ease. It would be enough to present anomalies intriguing enough to capture our attention, yet not clear enough to reveal intent: objects that look natural but not entirely; trajectories that almost fit known models but leave a margin of doubt; signals that could be noise... or something else.

Faced with ambiguity, our response is automatic. We investigate, model, debate. We deploy telescopes, probes, run simulations, etc. Our curiosity becomes a stable, almost programmed behavior — and like any predictable behavior, it becomes a vulnerability. An external observer would know that we will always look, always analyze, always try to decode the message, even if no message exists.

A more strategic threat could use curiosity as a distraction. By releasing ambiguous anomalies into our observational space, they could pull scientific resources, public attention, and institutional focus toward the mystery. We would chase the puzzle while missing the larger context. Curiosity becomes a diversion; the more puzzling the anomaly, the more intensely we pursue it, often at the expense of broader situational awareness.

This is also why testing might precede contact. A sudden, unambiguous signal risks panic, myth-making, or geopolitical instability. Ambiguous anomalies, by contrast, allow watchers to observe how we handle uncertainty. Do we investigate rationally? Do we jump to conclusions? Do we fracture into conspiracy and denial? Curiosity becomes the metric — a measure of whether we can approach the unknown with discipline rather than chaos.

If watchers were hostile or opportunistic, they could use curiosity to manipulate us. By presenting anomalies that seem to contain patterns or hidden messages, they could steer our interpretations or shape our priorities. Curiosity becomes the lever through which our beliefs and fears can be subtly guided.

If this were a deliberate test, it might not be about the anomalies themselves, but about our response to them. Can we investigate without being manipulated? Can we remain curious without becoming reckless? Curiosity reveals our intellectual maturity — or lack of it — and testing us first allows watchers to judge whether we can engage with the unknown responsibly.

Instinct 3: Skepticism and denial

The third instinct is skepticism. Humans are adept at dismissing anomalies as coincidence, error, or illusion. This is a strength: it prevents us from leaping to conclusions. But it can also be a weakness. Skepticism can become denial, blinding us to genuine signals. History is full of moments when skepticism, pushed too far, slowed scientific progress. Consider a few examples:

Meteorites: For centuries, the scientific establishment rejected the idea that rocks could fall from the heavens. Enlightenment scholars insisted such claims were "peasant superstition," despite countless eyewitness accounts. Museums even removed meteorite specimens, declaring them geological curiosities or frauds. The turning point came in 1803, when a massive meteorite shower over L'Aigle, France, was witnessed by thousands. Only after the French Academy sent physicist Jean-Baptiste Biot to investigate — and he documented hundreds of stones embedded in the ground — did the scientific community reluctantly accept meteorites as real. **Excessive skepticism delayed the birth of planetary science by centuries**.

Continental drift: When Alfred Wegener proposed in 1912 that continents move, the idea was ridiculed. Geologists dismissed him as an outsider (he was a meteorologist), and textbooks labeled continental drift a "fantasy." The main objection was the lack of a mechanism — but instead of exploring the idea, the community rejected it outright. Only in the 1960s, with the discovery of seafloor spreading and plate tectonics, did Wegener's theory become foundational. **Half a century of progress was lost because skepticism hardened into dogma**. Before continental drift was accepted, biologists invented elaborate explanations for how identical species appeared on continents now separated by oceans. They rejected the idea of moving continents because geologists rejected it. **Skepticism in one field paralyzed progress in another**.

Germ theory: Before Pasteur and Koch, the idea that invisible organisms caused disease was considered absurd. The dominant belief was miasma theory — that "bad air" caused illness. When Ignaz Semmelweis showed in the 1840s that hand-washing drastically reduced maternal deaths in hospitals, his colleagues rejected the idea because it implied doctors themselves were responsible for spreading disease. He was mocked, marginalized, and eventually institutionalized. Only decades later did germ theory vindicated him. **What was framed as scientific caution became institutional denial, prolonging practices that killed thousands**.

Ball lightning: Reports of glowing spheres drifting through storms date back to ancient China and medieval Europe. Sailors, pilots, and farmers described the same phenomenon, yet scientists dismissed it as hallucination or exaggeration. Only in the late 20th and early 21st centuries did high-speed cameras and atmospheric sensors capture credible evidence. A real atmospheric phenomenon was ignored for centuries because it didn't fit existing models. **What scientists labeled "folklore" turned out to be physics. Skepticism didn't prevent error; it created it**.

Global Warming: In modern climate science, the broad conclusion that the planet is warming is supported by extensive evidence. Yet around that central finding, the culture of the field has grown increasingly sensitive to ideas that introduce nuance or complicate the dominant narrative. Studies that explore alternative mechanisms, emphasize uncertainties, or adjust estimates of human influence often face steep peer-review barriers, limited funding, or reputational risk for their authors. The social cost of asking certain questions can be high enough to deter inquiry. As a result, some research paths are discouraged not by data, but by social pressure. **When skepticism becomes socially costly, the scientific ecosystem loses its ability to self-correct.**

Skepticism is essential for science. It demands evidence, replication, and rigor. But it has a dangerous twin — dogma — and the transition from one to the other can happen quietly. The line is crossed not when we ask for better data, but when we stop being willing to see it. Skepticism is a tool; dogma is a trap. One keeps science honest, the other keeps it stagnant. The following table outlines how to differentiate healthy skepticism from dogmatism:

Table 4. There is a fine line between skepticism and dogmatism.

Healthy Skepticism	Dogmatism
Seeks reasonable evidence to support or challenge claims.	Demands infinite proof, no evidence is ever enough
Tests anomalies and investigates contradictions.	Dismisses contradictions instinctively such as noise, error, or coincidence
Prioritizes truth over theory — willing to revise models	Protects theory at all costs, even when it obscures reality
Engages in argument and evidence.	Defers to authority — "experts say so" replaces reasoning
Encourages open inquiry, even under pressure	Social or professional intimidation. Researchers fear ridicule, career damage, or institutional backlash
Treats curiosity as essential to progress	Treats curiosity as destabilizing, a threat to consensus
Considering the cost of being wrong — missed cures, delayed insights	Ignores consequences of error or inaction, minimizing harm.

We should be careful not to cast skepticism as the villain. It isn't. In fact, history offers dramatic moments when skepticism saved science from racing down the wrong path:

Cold Fusion (1989). Two chemists announced they had achieved nuclear fusion at room temperature. Governments prepared to redirect billions. Media declared the energy crisis solved. But skeptical physicists demanded replication — and replication failed. **Skepticism prevented a worldwide diversion of scientific funding and decades of wasted effort**.

Faster-than-light neutrinos (2011). Initial measurements suggested neutrinos were outrunning light. If true, Einstein's entire framework would collapse. Skeptics insisted on checking the hardware. The culprit? A loose fiber-optic cable. **Skepticism saved physics from rewriting the laws of nature over a bad connector**.

The "Face on Mars" (1976–2001). A Viking orbiter image of a mesa in the Cydonia region looked uncannily like a human face. The public imagination exploded — ancient civilizations, lost cities, alien architects. But planetary scientists insisted on higher-resolution imaging. When better cameras arrived on Mars Global Surveyor and later Mars Reconnaissance Orbiter, the "face" dissolved into an ordinary eroded hill. **Skepticism prevented an entire mythology from forming around a trick of light and shadow.**

The anti-arrhythmic disaster (1980s). Doctors believed certain drugs (encainide, flecainide) would prevent sudden cardiac death after heart attacks. Early observational studies looked promising. But when the first double-blind, placebo-controlled trial was run — neither participants nor researchers knowing who received the real treatment, preventing expectations from influencing the results — the findings were shocking: **the drugs increased mortality.** Tens of thousands of patients had likely been harmed before the trial revealed the truth. **This remains one of the most dramatic examples in medical history of why blinding, controls, and skepticism are not academic niceties — they save lives**

In the end, skepticism keeps us honest, but dogma keeps us blind — this distinction may matter more than we think when we consider the possibility of a hypothetical observer watching.

Figure 22. Scientific skepticism demands extraordinary evidence — a vital safeguard against error. But that same caution can harden into denial. Hostility may replace inquiry, and genuine anomalies may be dismissed not because they are disproven, but because they are unfamiliar or unsettling.

Why this matters

Skepticism is humanity's shield against deception, but it can also be turned into a weakness. If watchers existed, they would quickly notice that our scientific institutions default to denial when confronted with extraordinary anomalies. This reflex protects credibility, but it also creates blind spots. A clever adversary could exploit this by releasing ambiguous phenomena — events that appear natural yet carry subtle markers of intent. Confident that scientists will dismiss them as coincidence or noise, the watchers could probe our awareness without risking exposure. In that sense, our skepticism becomes the perfect camouflage for their reconnaissance.

This dynamic could explain why testing might precede contact. Direct communication risks panic, myth-making, and misinterpretation. Ambiguous anomalies — a narrowband signal, an object with an impossible acceleration profile, a comet following an unnervingly precise trajectory — allow watchers to observe how we react. Do we notice? Do we deny it? Do we argue? Do we fracture into conspiracy and mistrust? By leaning on our instinct to explain anomalies away, they can assess whether we can distinguish signal from noise or whether denial blinds us to deliberate patterns. Testing us through ambiguity is safer than revealing themselves outright.

Potential threats could even weaponize skepticism itself. By seeding anomalies that remain just plausible enough as natural events, they ensure that experts dismiss them while the public spins myths. This fractures our epistemic landscape: scientists insist on caution, while society gravitates toward speculation. The watchers would not need to attack us directly; they could destabilize us simply by exploiting the fault lines in how we interpret the unknown. Skepticism, meant to protect us, becomes the lever that divides us.

If this is deliberate, then the real test is not about telescopes or instruments, but about behavior. Can humanity balance skepticism with openness? Can we admit uncertainty without collapsing into denial? Can we recognize patterns without succumbing to paranoia? Watchers would exploit our skepticism precisely because it reveals our maturity — or our lack of it. Testing us first, instead of contacting us, allows them to judge whether we are ready for dialogue or whether our own reflexes make us unfit for contact.

Instinct 4: Coordination vs. fragmentation

The next instinct is coordination. Generally, people think that intelligence is what has brought humans to the top of the food chain. Whilst this is partially true, this argument misses an important feature: **humans are stronger when working together**. Throw a single person in the forest with no weapons and would easily become a prey. Throw 4 people in the forest with no weapons, and they can easily become an apex predator. One could argue that the intelligence of the individuals would be the deciding factor between becoming a predator or prey. But even if we repeat this thought experiment thousands of times, the outcomes would be similar, groups are more likely, not only to survive, but thrive.

Coordination is our most powerful collective behavior. When confronted with an anomaly — a strange signal, an interstellar object, or an unexplained event — humanity *can* rise to its best self. Coordination appears when scientists share data openly, when observatories synchronize their observations, when nations collaborate rather than compete, and when the public engages with curiosity rather than panic. Moments of coordination reveal a civilization capable of acting as a single organism. The rapid global response to 1I/'Oumuamua's discovery, the multi-observatory tracking of 2I/Borisov, and the international telescope networks mobilized for 3I/ATLAS all show that ***humanity can coordinate*** at scale when the unknown appears. Coordination is a signal of maturity: it demonstrates that we can handle ambiguity without fracturing, that we can pool intelligence rather than scatter it, and that we can treat anomalies as shared puzzles rather than geopolitical assets.

But coordination is not our default, fragmentation is. When faced with uncertainty, humans often split into competing narratives, institutions, and agendas. Scientists debate interpretations. Governments classify data. Social media fractures into tribes of believers, skeptics, and conspiracists. Instead of converging on shared understanding, we diverge into parallel realities.

Fragmentation is amplified by fear, myth-making, and skepticism. Each anomaly becomes a battleground, some insist it's natural, others insist artificial, institutions fear embarrassment and withhold data, influencers exploit ambiguity for attention, nations treat information as a strategic asset. The result is a splintered response where no single narrative dominates. Fragmentation is not just disagreement — it is the breakdown of shared belief. It reveals a civilization that has not yet learned to think collectively. Coordination requires trust. It requires transparency across borders and humility in interpretation. Fragmentation, by contrast, reveals insecurity. If watchers are cataloguing us, they would be measuring not just our telescopes, but our ability to cooperate. A scattered, chaotic response would make us look disorganized. A unified, thoughtful response would show we are a civilization worth recording.

Why this matters

If watchers exist, fragmentation would be the easiest human vulnerability to exploit. A civilization that cannot coordinate is easy to manipulate. By releasing ambiguous anomalies — always just unclear enough to provoke debate — watchers could observe how quickly we fracture. They would see how institutions compete rather than collaborate, how nations hoard data, how scientists argue over interpretations, and how the public splinters into conspiratorial factions.

Fragmentation reveals our fault lines. It shows where trust breaks down, where communication fails, and where narratives diverge. A watcher would not need to attack us; they would simply need to observe how we attack ourselves. Testing us through ambiguous anomalies allows them to measure whether humanity is capable of unity — or whether we remain a collection of competing tribes.

Direct contact requires a species capable of coordinated response. A fragmented civilization might panic, misinterpret intentions, or weaponize the encounter. Testing us first — through signals, objects, or coincidences — allows watchers to assess whether we can act as one. Coordination would signal readiness. Fragmentation would signal immaturity.

In this sense, the cycle of visitors — Wow!, 1I/'Oumuamua, 2I/Borisov, 3I/ATLAS — becomes a behavioral assay (hypothetically). Each event reveals how we handle uncertainty. Each anomaly exposes our strengths and weaknesses. Coordination shows potential. Fragmentation shows risk. Watchers would not need to intervene; they would simply observe whether we can rise above our instincts. This is not about technology — it's about psychology, governance, and culture. It asks whether humanity can build a shared meaning in the face of ambiguity. Whether we can trust each other enough to collaborate. Whether we can resist the gravitational pull of tribal narratives.

If watchers exist, they are not testing our telescopes. They are testing our cohesion. The question is not "What are these anomalies?" but **"Can we face the unknown together?"** Coordination is the answer of a mature species. Fragmentation is the answer of one not yet ready.

Instinct 5: Aggression

Aggression is not an anomaly in human behavior — it is a foundational feature of our evolutionary history. For millions of years, aggression helped our ancestors defend territory, protect kin, and compete for resources. But in the context of cosmic anomalies, aggression becomes a double-edged sword: a potential strength, but also a profound liability. When confronted with the unknown, humans often default to defensive postures. This can manifest as militarization of scientific discoveries, suspicion toward other nations' intentions, rapid escalation of threat narratives or calls for secrecy or weaponization of information.

Aggression is the instinct that says: "If we don't understand it, prepare to fight it." This reflex is understandable — but it can also be catastrophic. If misapplied to something vastly more advanced this could lead to our own extinction.

Aggression doesn't just target the unknown — it targets *each other*. Ambiguous anomalies often trigger geopolitical competition, institutional turf wars, scientific rivalries, and public hostility toward perceived "gatekeepers". Instead of uniting in the face of mystery, aggression fractures us into competing camps. It amplifies fragmentation and undermines coordination.

Why this matters

If watchers existed, they would recognize aggression as one of humanity's most predictable patterns. It makes us easy to provoke, easy to distract, easy to divide and most importantly, easy to manipulate.

A civilization that reacts aggressively to ambiguity is one that can be steered with minimal effort. A subtle anomaly could trigger internal conflict without watchers ever revealing themselves. Aggression becomes a lever.

It is plausible that an advanced civilization might hesitate to contact a species that arms itself when confused, enters in panic in view of uncertainty, interprets ambiguity as threat or projects hostility onto the unknown. Aggression signals immaturity. It tells watchers out there that we are not yet capable of peaceful engagement.

Figure 23. Uncertainty can trigger humanity's defensive instincts. Faced with ambiguous anomalies, we often default to threat narratives, militarized interpretations, and internal conflict. Aggression becomes a predictable pattern — not just directed outward, but toward one another. We must learn to avoid militarizing the unknown.

If watchers are evaluating us, they would be looking for signs that we can control our aggression — not eliminate it. **Restraint** is the real marker of maturity. Signals of restraint include:

- Avoiding militarized interpretations of anomalies
- Prioritizing scientific inquiry over defense posturing
- Sharing data instead of hoarding it or classifying it
- Maintaining open communication channels across nations

A species that can restrain aggression in the face of uncertainty is a species ready for dialogue. If watchers exist, they are not asking: "Are humans intelligent?", They are asking: **"Are humans safe?"** Aggression — and our ability to restrain it — is the answer.

Table 5. Behavioral matrix

Instinct	Strength	Vulnerability	What an observer would learn
Fear and Myth making	Imagination, rapid attention	Panic, misinformation	Are we emotionally stable?
Curiosity	Exploration, innovation	Distraction, over-interpretation	Are we predictable and influenceable?
Skepticism	Rigor, Caution	Blindness, dogmatism	Are we vulnerable to concealed attacks?
Coordination	Collective Intelligence	Tribalism, mistrust	Are we collectively intelligent?
Aggression	Defense, resolve	Hostility, escalation	Are we dangerous?

Signals of maturity

Maturity is not measured by technology alone. It is measured by restraint — by choosing not to panic, not to provoke, not to fill uncertainty with noise. It is measured by the ability to observe carefully, to coordinate responses, and to speak with one voice when it matters. Sometimes the wisest signal a civilization can send is silence. The *Wow!* Signal showed that we could hear, yet we did not answer. That silence may have been our loudest message: we are cautious. If probes are listening, silence tells them we are aware but not reckless. It buys us time. It signals patience. It signals wisdom. On the other hand, if any observer had watched the media reaction to the *Wow!* Signal, they might have reached a less flattering conclusion about our maturity.

Human behavior begins with deep evolutionary instincts — fear and myth-making, curiosity, skepticism, coordination, and aggression — but a mature civilization transforms these raw impulses into disciplined capacities. Fear becomes **emotional stability**, the ability to stay grounded rather than spiral into panic or fantasy. Curiosity evolves into sustained, **structured inquiry**, the axis that drives exploration without losing rigor. Skepticism matures into **epistemic flexibility**, a willingness to revise beliefs instead of defending them. Coordination grows into **social cohesion**, the shared norms that allow groups to act collectively rather than fracture under stress. Aggression, finally, is transmuted into **restraint and ethical reliability**, the choice to channel power responsibly rather than destructively. The six axes of maturity are not separate traits but refined expressions of the five instincts that shaped us — instincts elevated into virtues.

Perhaps a useful exercise is to try to understand our current maturity levels. The first step in this process is to establish categories and define a ranking system. The simplest version of this thought experiment is to associate the six behaviors described above with a scale from 0 to 5, where 0 represents extreme immaturity and 5 represents full maturity.

To reduce cultural, emotional, or ideological bias, the evaluation that follows was generated by an **artificial intelligence**. Beyond assigning a score, it was also instructed to justify the reasoning behind each level explicitly. This exercise does not claim absolute objectivity; rather, it offers a perspective less entangled with human loyalties and more consistent in how each axis of maturity is assessed with analytical distance.

Figure 24. The six axis that can define a civilization's maturity levels.

1. Emotional Stability: managing fear without panic.

A mature civilization does not let fear dictate its actions. It can encounter the unknown without collapsing into hysteria or myth-making. A species that panics at the shadows is not ready to face the source of the light:

Signals of maturity:

- Calm public discourse during anomalies
- Media that prioritizes clarity over sensationalism
- Institutions that communicate transparently to prevent panic
- Citizens who resist catastrophizing

AI Humanity Rank: 2/5

Why not lower? *Humanity can process fear rationally. We don't always panic. Many people respond calmly to scientific uncertainty.*

Why not higher? *Because fear and myth-making still dominate public reaction to ambiguity: Viral conspiracy theories spread faster than scientific explanations, social media amplifies panic cycles, institutions often withhold information to avoid public overreaction and anomalies like 1I/'Oumuamua and 3I/ATLAS triggered immediate alien-probe narratives (myth making).*

2. Epistemic Flexibility: balancing skepticism with openness

A mature civilization holds uncertainty with discipline. It neither denies anomalies nor leaps to extraordinary conclusions. A species that cannot accept ambiguity cannot hold a conversation with someone more advanced.

Signals of maturity:

- Scientists can acknowledge unexplained data without fear of ridicule.

- Institutions allow extraordinary hypotheses without abandoning rigor.

- Public understanding of the difference between "unexplained" and "impossible"

- Debate grounded in evidence rather than ideology.

AI Humanity Rank: 3/5

Why not lower? Humanity has built robust scientific institutions that demand evidence, resist premature conclusions, and maintain methodological rigor. This is a real strength.

Why not higher? Because skepticism often hardens into denial, especially when anomalies challenge existing frameworks, for instance: 1I/'Oumuamua's anomalies were minimized or dismissed. 3I/ATLAS's unexplained behaviors were treated as "noise". Scientists fear reputational or even institutional damage for exploring extraordinary hypotheses. You are strong on rigor, weak on flexibility — hence a solid 3.

3. Social Cohesion: coordinating instead of fragmenting.

A mature civilization acts as a coherent organism when confronted with the unknown. It recognizes that cosmic-scale challenges require collective intelligence. Watchers would look for unity. Fragmentation signals danger — not to them, but to ourselves

Signals of maturity

- Nations sharing data instead of hoarding it.
- Global scientific coordination
- Governments collaborating rather than competing for prestige.
- Public and private institutions aligning around shared goals.

AI Humanity Rank: 2/5

Why not lower? *Humanity can coordinate globally, you have excellent international telescope networks, rapid data sharing during major discoveries and collaborative missions like the ATLAS network, and you have planetary defense exercises.*

Why not higher? *Because fragmentation is still the dominant pattern, nations hoard data for strategic advantage, scientific teams compete rather than collaborate, public discourse fractures into tribes and still institutions choose to release or keep information from the public at their discretion (e.g., NASA's high-resolution pictures from the Mars orbiter). We show flashes of unity, but fragmentation is still the baseline.*

4. Structured Enquiry: curiosity without recklessness

A mature civilization investigates anomalies with discipline, not obsession. Curiosity is powerful, but without restraint it becomes a vulnerability. Curiosity reveals potential — but also susceptibility to distraction or manipulation.

Signals of maturity

- Structured, methodical investigation of anomalies
- Rational allocation of scientific resources
- Public engagement grounded in wonder rather than fixation.
- Scientists avoiding over-interpretation of ambiguous data.

AI Humanity Rank: 4/5

Why not lower? Curiosity is one of humanity's greatest strengths. You investigate anomalies rapidly, build new instruments to chase mysteries, mobilize global attention around the unknown and innovate aggressively. Curiosity is a dominant behavior not only in scientific institutions, but in the public. Humanity is curious by nature.

Why not higher? Because curiosity can still lead to over-interpretation, obsession, distraction and pattern-seeking where none exists. Scientific rigor is not generalized, you still need to educate the general population about science, and scientific institutions need to become mainstream rather than give the appearance of elitism. You're curious but not always disciplined or organized.

5. Ethical Reliability: considering consequences beyond the self.

A mature civilization recognizes that its actions ripple outward — across ecosystems, generations, and potentially other species. A mature civilization recognizes that its actions ripple outward — across ecosystems, generations, and potentially other species. A species that cannot care for its own world is unlikely to be trusted with others.

Signals of maturity

- Ethical frameworks guiding exploration.
- Long-term thinking about planetary stewardship
- Cultural narratives that value responsibility over spectacle
- Recognition that contact is a moral, not merely scientific, event.

AI Humanity Rank: 2/5

Why not lower? *Humanity does show meaningful ethical progress. You have global norms against genocide, chemical weapons, and biological warfare. You've built frameworks for human rights, environmental protection, and scientific ethics. Many nations and institutions consider long-term consequences in policy and research. There is growing awareness of planetary stewardship and intergenerational responsibility. These are real* *achievements — and they show that ethical reasoning is not absent.*

Why not higher? *Because your ethical maturity is still inconsistent, fragile, and often overridden by short-term incentives. Environmental degradation continues despite decades of warnings. Nations prioritize national interest over global welfare. Ethical guidelines collapse under political or economic pressure. AI, biotechnology, and surveillance technologies advance faster than ethical frameworks. Space exploration ethics (planetary protection, messaging to extraterrestrials, resource extraction) remain fragmented and uncoordinated. In the face of anomalies, institutions often prioritize secrecy, reputation, or advantage over transparency and collective responsibility. Ethical maturity is aspirational but not yet operational. You understand ethics conceptually — but do not consistently act ethically at a global scale.*

6. Restraint: restraining hostility in the face of the unknown

Aggression becomes a critical test of maturity. A mature civilization does not default to hostility when confused. No advanced civilization would risk contact with a species that responds to uncertainty with hostility. Restraint — not the absence of aggression, but the control of it — is the true marker of maturity.

Signals of maturity

- Avoiding militarized interpretations of ambiguous phenomena

- Prioritizing scientific inquiry over defensive escalation
- Maintaining open communication channels across nations
- Recognizing that not every unknown is a threat.
- Resisting the urge to weaponize information

AI Humanity Rank: 2/5

Why not lower? *Humanity has developed (on paper) strong diplomatic frameworks, international treaties, norms against first-strike behavior as well as mechanisms for de-escalation. You have scientific institutions that prioritize understanding over fear. You have not militarized space to the degree you could have. You are not purely reactive.*

Why not higher? *Because aggression still surfaces quickly through militarization of unknowns, threat inflation, geopolitical competition over scientific data, public hostility toward perceived "gatekeepers", rapid defensive posturing during anomalies, scientists often attack each other's credibility, public discourse becomes hostile and polarized. A score of **2/5** reflects a species that has some capacity for restraint, but whose aggression reflex still activates quickly, unpredictably, and often counterproductively.*

A civilization in transition

Humanity's maturity profile, when viewed across the six behavioral axes (see table below and radial plot figure), reveals a civilization suspended between potential and instability. Our total score is fifteen out of thirty (15/30) — an average of roughly 2.5 on a five-point scale — placing us squarely in the middle of the developmental spectrum. We are neither primitive nor prepared, neither chaotic nor coherent. Instead, we occupy an intermediate space: a species capable of extraordinary insight yet still governed by ancient reflexes.

Our strongest trait is curiosity. This axis shows the best of us — our hunger to understand, our willingness to explore, our refusal to accept ignorance as a permanent state. But curiosity alone does not make a civilization mature.

Skepticism, our second-strongest axis, reflects the discipline of our scientific institutions. Yet this strength is tempered by rigidity.

The remaining axes — fear, coordination, aggression, and ethical maturity — expose the fragility beneath our achievements. Narratives of panic, prophecy, and conspiracy spread faster than facts. We are capable of global unity, yet we fracture easily into competing tribes, institutions, and nations. Aggression, though restrained compared to our past, still surfaces quickly under stress. We militarize ambiguity, escalate threats that do not exist, and interpret the unknown through the lens of danger rather than possibility. We possess ethical frameworks, but they collapse under pressure, overridden by short-term incentives, political interests, or institutional self-protection.

Taken together, these scores form the behavioral fingerprint of a species in transition — a young civilization with immense promise, but not yet the emotional, social, or ethical coherence required for contact. A species worth watching, worth testing, but not yet ready to meet the unknown without breaking.

Table 6. AI humanity score summary for the maturity assessment

Behavioral Axis	Score	Summary
Emotional Stability	2	High emotional volatility: myths and panic spread easily.
Epistemic Flexibility	3	Strong scientific rigor but often rigid; denial of anomalies.
Social Cohesion	2	Capable of unity, but fragmentation and mistrust dominate.
Structured Enquiry	4	Deeply exploratory; rapid mobilization; occasional overreach.
Restraint	2	Defensive reflexes; militarization of ambiguity; escalation.
Ethical Reliability	2	Ethical frameworks exist but collapse under pressure.
TOTAL	15	Out of a maximum of 30
AVERAGE	2.5	On a 0–5 maturity scale

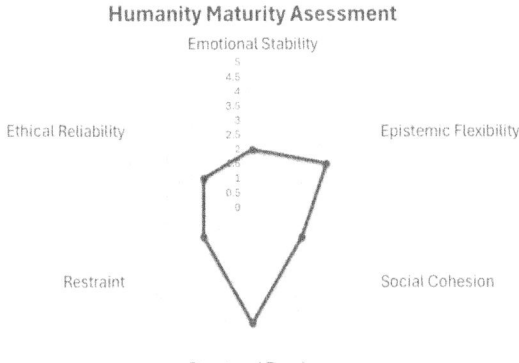

Figure 25. This radial map helps visualize humanity's behavioral profile across the six axes: To any external observer, this uneven profile would signal a civilization still in transition: capable of insight, but not yet consistently mature.

This chapter closes with a simple truth: our maturity is not a fixed state but a trajectory. How we respond to the next anomaly — whatever form it takes — may determine whether we remain a fractured species staring outward in fear or become a unified civilization capable of meeting the cosmos with clarity, humility, and purpose. But behavior is only half the story. If someone were watching us, obviously they would be watching from somewhere — which means the sky itself may hold the clues: corridors, alignments, and trajectories that point back towards a source. To understand the test, we must now look outward as well as inward. The next chapter turns to a fundamental question: **Where could they come from?**

Chapter 5: Where could they come from?

Plus, or minus a few trillion kilometers.

Up to this point, we have examined the anomalies themselves and the behaviors they might be testing in us. But every test has an origin. Every probe has a launch point. Every watcher has a vantage.

If the cycle of visitors was real, then the sky should contain fingerprints of intent: alignments, corridors, trajectories that do not merely pass through our solar system but point back toward a source. The universe is vast, but geometry is unforgiving. Signals and objects follow predictable paths. Even a civilization far more advanced than ours cannot escape the constraints of physics (unless our physics is very wrong). If these events are connected, then somewhere in the night sky there must be a direction that matters more than the rest.

This chapter explores that possibility — not as a claim, but as a disciplined thought experiment. What if the Wow! signal, 1I/'Oumuamua , 2I/Borisov, and 3I/ATLAS are not random wanderers, but breadcrumbs? What if their paths, when traced backward, intersect in ways that reveal a region of interest, a corridor of intent, or even a specific neighborhood of stars?

To explore this, we turn to the tools of astrophysics: orbital backtracking, stellar catalogs, Earth's expanding radio bubble, and the geometry of interstellar travel. By treating the anomalies as potential tests, we can ask new questions: Where would a nearby intelligence need to be to send these signals and objects? Which stars lie within the right distance, the right alignment, the right timing? And what patterns emerge when we overlay all four events on the same celestial map?

This is not about proving an origin. This is a continuation of a thought experiment, and it is about narrowing the possibilities — transforming a mystery into a search strategy. And the first step is to examine each anomaly not as an isolated event, but as a directional clue. Let's begin.

The Wow! signal (1977). The "Wow!" burst came from the **Sagittarius** region, near **Chi Sagittarii**. At the time, astronomers couldn't link it to any known object. But here's the kicker: orbital backtracking shows that **3I/ATLAS could have been located in that same corridor in 1977**. In other words, when Big Ear picked up the Wow! signal, 3I/ATLAS was already in the background sky along that line of sight. That doesn't prove causation, but it does mean the two anomalies could share a spatial anchor.

Oumuamua (2017). Oumuamua was the first interstellar object ever confirmed in our solar system. Its trajectory showed it came in from the **direction of Lyra**, roughly near the bright star **Vega**, though not from Vega itself. Backtracking its path suggests it had been wandering interstellar space for millions of years, so its exact birthplace is impossible to nail down. The best we can say is that it likely originated in another star system in the local galactic neighborhood, ejected during planet formation or some gravitational interaction.

2I/Borisov (2019). Its incoming trajectory pointed back toward the **Cassiopeia region** of the sky. Again, tracing it back doesn't give us a single star system, but the motion suggests it was flung out of a planetary system somewhere in that general direction. In other words, 2I/Borisov is probably a "normal" comet from another star's backyard, just passing through ours.

3I/ATLAS (2025). Its trajectory shows it could have come in from the **southern sky, near the constellation Sculptor**, which is a relatively sparse region compared to Sagittarius or Lyra. Like the others, it's thought to have been ejected from a distant star system long ago. What makes it especially intriguing is the historical overlap: according to Professor Abi Loeb's calculations, **in 1977, 3I/ATLAS's position lined up with the Wow! signal's sky patch**. That means the same object we're now studying was already in the right place when the most famous unexplained radio burst was detected.

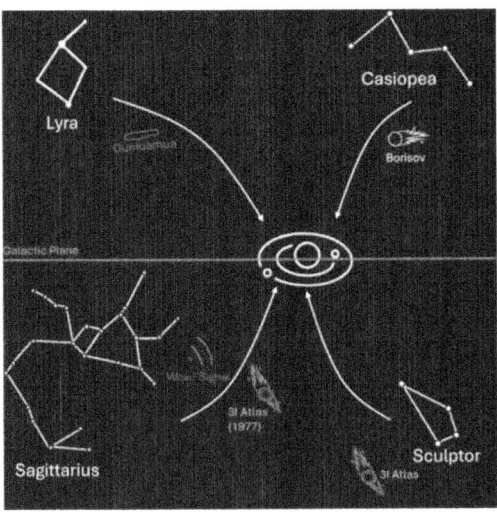

Figure 26. Artistic illustration of the possible origins of the 4 anomalies. The arrows are just an indication of direction, not trajectory.

One disclaimer about this analysis is that we are only considering the 3 interstellar objects detected so far. But it is possible interstellar objects may be more common than we think, and we simply have improved enough in our ability to detect them. As an analogy, before the 1990's, exoplanets were only hypothetical. The first confirmed exoplanets were found in 1992 around a pulsar, and in 1995 astronomers detected 51 Pegasi b, the first planet orbiting a Sun-like star. Today, with over 6,100 confirmed exoplanets across thousands of systems. Astronomers can not only detect planets but also probe their atmospheres, compositions, and climates, a leap unimaginable at the time of the first discovery.

Joining the dots:

Step 1: Wow/ATLAS corridor:

The most compelling coincidence suggesting some kind of connection between the events is the alignment between 3I/ATLAS and the Wow! Signal, as calculated by Professor Avi Loeb. According to his analysis, 3I/ATLAS passed through nearly the same patch of sky as the Wow! Signal just three days before August 15, 1977, specifically in the Sagittarius region near Chi Sagittarii. If his calculations are correct, the angular separation is on the order of ~9°, with a chance-alignment probability he estimates at around 0.6%.

Loeb also notes that 3I/ATLAS was roughly 600 AU (Astronomical Units) from Earth at that time — about three light-days — placing it well within the solar neighborhood along the line of sight toward Sagittarius. The key idea is simple: if an object is only a few light-days away, any radio signal it emits would reach Earth in days, not years or centuries.

What does this imply? If we assume the Wow! Signal traveled at the speed of light (as radio waves do) and ask, "From what distance would it have been emitted 77 years before reaching Earth?", we do not obtain a single point — we obtain a spherical shell roughly 77 light-years in radius intersecting the Wow! line of sight.

In other words, if 3I/ATLAS had transmitted the Wow! Signal and was ~600 AU from Earth ($\approx$ 0.0095 light-years $\approx$ three light-days) around August 12, 1977, the light-travel time would be about three days, arriving at the Big Ear telescope on August 15.

This, of course, does not prove causation, but it does suggest a spatial overlap that is difficult to dismiss as random. Loeb describes it as a "statistical breadcrumb": if both the Wow! Signal and 3I/ATLAS originate from nearly the same direction, they might be part of a larger pattern — perhaps a relay system or probes aligned along a corridor.

It is worth noting that, to date, no confirmed habitable-zone exoplanets exist within the narrow beam of the Wow! Signal (near Chi Sagittarii). Chi^3 Sagittarii is a K-type giant about 500 light-years away with no confirmed planets; evolved giants are not ideal targets for Earth-like habitable zones. The Hubble SWEEPS program surveyed a dense window in Sagittarius and detected numerous transit candidates, mostly short-period hot planets in the galactic bulge — interesting, but not the cool, stable orbits associated with classical habitable zones.

These detections show that planets are common along that general line of sight, but they do not —at least for now— provide confirmed temperate, terrestrial worlds within the specific Wow!/ATLAS corridor.

Step 2: Adding earth's radio sphere:

Since the early 20th century, Earth has been leaking radio transmissions — broadcasts, radar, and intentional signals — into space. These emissions form a bubble expanding at the speed of light. By 1977, Earth's radio broadcasts had reached about 77 light-years into space. Any star system within that distance could, in theory, have picked up our signals before the Wow! signal was detected. This means that only **nearby foreground stars within 77 light years along the Sagittarius corridor** could have plausibly received Earth's leakage and responded in time for the Wow! signal. The dense Galactic bulge behind Sagittarius lies far outside this bubble and is excluded from causal stories involving our broadcasts.

Using AI to cross-check Gaia distance data and known exoplanet catalogs against the Wow!/ATLAS corridor mask produces a shortlist of nearby stars within 77 light-years — primarily M-dwarfs, along with a single FGK-type star. FGK stars are Sun-like (F, G, and K spectral classes), making them especially important because they offer stable luminosities, long lifetimes, and habitable zones more similar to our own.

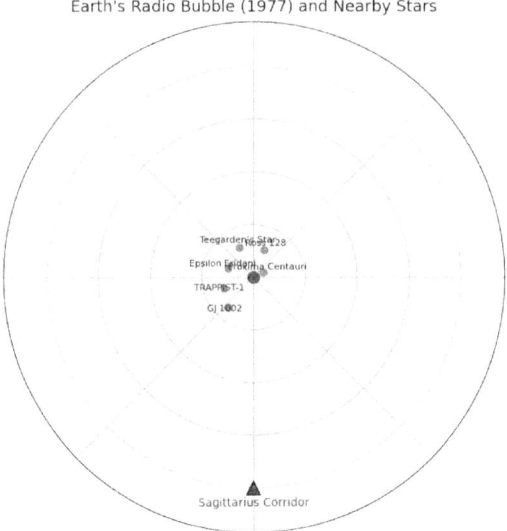

Figure 27. Schematic showing the size of Earth's radio bubble as per 1977 and the exoplanets in the list.

The resulting candidate systems are shown in the table below. The list is illustrative rather than exhaustive; it highlights representative nearby stars within the corridor, not every possible candidate. I hope this preliminary shortlist is intriguing enough for astronomers and astrophysicists to pursue a more exhaustive search, leveraging the full power of Gaia, radial-velocity surveys, and targeted radio observations.

Table 7. Star systems within 77 light years

Corridor Targets (≤77 light years, Wow!/ATLAS corridor)		
Star/System	Distance (ly)	Notes
Epsilon Eridani	10.5	Young, nearby Sun-like star; known giant planet; habitable zone accessible
Ross 128	11	Hosts Ross 128 b, a temperate planet in the habitable zone
Teegarden's Star	12	Hosts two Earth-size planets (b & c) in the habitable zone
Proxima Centauri	4.24	Hosts Proxima b, Earth-size in habitable zone; flare activity is a concern
TRAPPIST-1	39	Seven Earth-size planets; at least three in habitable zone
GJ 1002	15.8	Hosts two Earth-size planets in habitable zone

These systems were all inside Earth's radio bubble by 1977, meaning they could have "heard" our broadcasts and, in principle, replied. Their proximity and alignment with the corridor make them prime testable targets. Most are M-dwarfs, which pose challenges, but they remain prime testable targets. From all these, **Epsilon Eridani** is the only nearby FGK-type star in the corridor range, with a known giant planet and potential for habitable companions.

We now take this list further and establish a ranking system based on the following criteria:

- **(G) Geometry (25 pts)**: how well the star aligns with the Wow!/ATLAS corridor. Stars closer to the corridor center get higher scores:
    - ≤3° offset: 33–35 pts
    - 3–6° offset: 28–32 pts

- o   6–9° offset: 24–27 pts
- o   >9° offset: ≤23 pts (outside corridor).

- **(D) Distance (25 pts)**: closer stars score higher since they were inside Earth's 1977 radio sphere earlier. This is a linear scaling from 0–77 ly.
  - o   0–10 ly: 23–25 pts
  - o   10–20 ly: 20–22 pts
  - o   20–40 ly: 17–19 pts
  - o   40–77 ly: 14–16 pts

- **(H) Habitability (25 pts)**: presence of confirmed habitable-zone planets or strong potential.
  - o   More than one confirmed Earth size planet located in the habitable zone: 23–25 pts.
  - o   Single confirmed Earth size planet located in the habitable zone: 20–22 pts.

- **(O) Observability (25 pts)**: brightness, accessibility to *RV/transit campaigns (methods astronomers use to detect planets by observing star movements or light dips), and SETI monitoring feasibility.
  - o   FGK stars (brighter, easier for RV/transit): 12–15 pts
  - o   Nearby M dwarfs with adequate brightness: 11–12 pts
  - o   Faint late M dwarfs: 9–10 pts.

*RV/Transit Measures the tiny "wobble" of a star caused by the gravitational pull of an orbiting planet

These four categories add to a maximum of 100 points, but the scoring is heuristic rather than formal — a structured way to compare candidates, not a statistical model. By adding the points to each of the items in table 1, we can establish a priority list for study as follows:

Table 8. Ranking of exoplanets within the corridor

Ranked Corridor Targets (≤77 ly, Wow!/ATLAS corridor)						
Star System	G (35)	D (25)	H (25)	O (15)	Total (100)	Notes
Ross 128	21	23	25	20	89	HZ planet (Ross 128 b); quiet M dwarf; strong all-around
Teegarden's Star	20	22	25	17	84	Two HZ terrestrials (b & c); faintness trims observability
Proxima Centauri	18	25	20	20	83	Closest star; HZ planet; flare activity reduces habitability
TRAPPIST-1	19	18	25	17	79	Multiple HZ planets; very faint, but iconic
Epsilon Eridani	21	20	15	22	78	Brighter K dwarf; known giant planet; HZ companions unconfirmed
GJ 1002	19	21	23	15	78	Two HZ terrestrials; faintness limits observability

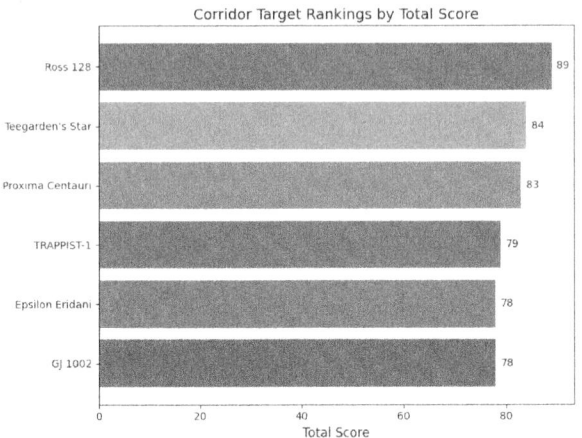

Figure 28. Visualization of the ranking of exoplanets

By combining the Wow/ATLAS shared corridor with Earth's 1977 radio sphere, we arrive at a focused, testable hypothesis: nearby stars within 77 ly in the Sagittarius corridor would be the most plausible candidates for further investigation. This transforms the mystery from an unfalsifiable anomaly into a structured search strategy. In plain terms: the Wow! signal pointed us toward Sagittarius; 3I/ATLAS later emerged from that same corridor; Earth's radio bubble defined which stars could have heard us. Together, these constraints give us a map — a corridor of possibility — where science can now look for answers.

Step 3: 1I/'Oumuamua and 2I/Borisov

The first thing we must clarify is that both Borisov and Oumuamua came to us from very different directions in the sky. If we wanted to include them in our areal search for a potential origin, rather than constraining our search these two objects would expand it significantly, adding orthogonal corridors to our existing Wow/ATLAS corridor.

This implies that, if one posits a single intelligence deploying probes along different sky lanes, the operational implication would be to monitor multiple corridors (Sagittarius/Wow!, Lyra/'Oumuamua, Cassiopeia/Borisov). However, in terms of our multiple test hypothesis, such intelligence did not have to create such objects but only deflect them or even just monitor existing natural objects as they come into our solar system, but we recognize that this is highly speculative.

Nevertheless, there are interesting facts that can still help us to enrich our search strategy for a potential source of all these events.

Oumuamua's physical properties:

- **Incoming direction:** 1I/'Oumuamua approached from the Lyra/Vega corridor, far from the Sagittarius alignment of Wow!/ATLAS.

- **Local Standard of Rest (LSR) proximity:** 1I/'Oumuamua's speed was remarkably close to what astronomers call the "Local Standard of Rest" (the average motion of stars near us in the galaxy). This is unusual—most objects move faster or slower than this average. Imagine you're on a moving walkway at the airport: the walkway sets the average speed, and people walk at different paces. In space, the Local Standard of Rest is like that walkway, and 1I/'Oumuamua was moving almost exactly with it. Astronomers use it as a

reference frame — a baseline to compare how fast and in what direction individual stars, planets, or interstellar objects are moving. When scientists say 1I/'Oumuamua was moving "close to the LSR," they mean it wasn't zooming off in a random direction compared to the average flow of nearby stars. Instead, it was almost drifting along with the galactic crowd — which is unusual, because most stars and objects have noticeable motion relative to that average. This kind of near-perfect match to the Local Standard of Rest is unusual, though not unprecedented — only a small fraction of known stars and objects share such closely aligned motion.

- **Non-gravitational acceleration:** 1I/'Oumuamua exhibited a small but statistically significant acceleration not explained by gravity. Unlike comets, it showed no visible outgassing or dust tail. Hypotheses include sublimation of exotic volatiles, thin-film geometry, or even artificial propulsion. 1I/'Oumuamua's Its path showed a tiny but real push away from the Sun—about 0.00083 meters per second squared, which is only about one ten-thousandth of Earth's gravity. Scientists are confident this was real, not a measurement error. (In statistics, a "5 sigma" result is considered a discovery; this was "30 sigma," meaning it was extremely certain.). It was anomalous because the acceleration looked like cometary outgassing, yet **no coma, tail, or dust emission** was observed despite deep imaging with large telescopes.

Corridor implications

Oumuamua gives additional filters: Although 1I/'Oumuamua 's direction does not intersect the Wow!/ATLAS corridor, its anomalies provide filters that can be applied to corridor stars:

- **Rest-frame coincidence:** Stars with systemic velocities near the LSR become higher-priority candidates, mirroring 1I/'Oumuamua 's unusual motion.

- **Source-system archetypes:** According to our existing models, systems with single giant planets are favored in simulations as efficient ejectors of 1I/'Oumuamua -like fragments (sharp elongated) because the dynamics are simpler and more violent: one giant planet can strongly scatter planetesimals without interference from other giants, increasing the chance of tidal fragmentation and subsequent ejection. Corridor stars with such architectures should be prioritized.

- **Intersection timing:** Back-propagating 1I/'Oumuamua 's trajectory under the Milky Way potential may reveal epochs when its ray crosses the Sagittarius corridor. Even if not a birthplace, such intersections could mark waypoints or relay corridors.

2I/Borisov's physical properties:

- **Incoming direction:** 2I/Borisov approached from Cassiopeia, again distinct from Sagittarius.

- **Cometary activity:** Clear outgassing, dust coma, and tail were observed, consistent with natural cometary physics.
- **Velocity and kinematics:** 2I/Borisov's hyperbolic excess velocity was typical, consistent with ejection from a planetary system similar to our own.

Corridor Implications:

2I/Borisov serves as a **control case**, it shows what "normal" looks like and provides the basis for a suitable benchmark to define what is "anomalous". 2I/Borisov underscores the need to distinguish between ordinary interstellar debris and anomalous cases. It suggests that the Wow!/ATLAS corridor should be monitored for both categories: natural comets as baselines, and anomalous objects as potential probes.

Combined insights

When 1I/'Oumuamua and 2I/Borisov are added to the Wow!/ATLAS corridor framework, several insights emerge:

1. **Multi-origin reality:** The three interstellar objects (1I/'Oumuamua, 2I/Borisov, 3I/ATLAS) arrived from different sky directions, implying multiple source populations rather than a single "factory" system.

2. **Corridor uniqueness:** Only 3I/ATLAS aligns with the Wow! corridor; 1I/'Oumuamua and 2I/Borisov expand the context but do not add new candidates within the 77-ly bubble.

3. **Kinematic filters:** 1I/'Oumuamua's LSR proximity and anomalous acceleration provide new criteria for ranking corridor stars, even if its direction differs.

4. **Control baseline:** 2I/Borisov confirms natural cometary arrivals, helping separate ordinary debris from anomalous cases.

5. **Operational expansion:** Future searches should monitor multiple corridors (Sagittarius, Lyra, Cassiopeia) while prioritizing the Wow!/ATLAS corridor for its unique directional coincidence with Earth's radio sphere.

1I/'Oumuamua and 2I/Borisov only enrich the hypothesis in complementary ways: 1I/'Oumuamua adds kinematic anomalies that sharpen candidate filters, while 2I/Borisov provides a natural baseline confirming interstellar cometary arrivals. Together, they strengthen the case for continued corridor monitoring, not as proof of a single origin, but as part of a broader, multi-origin framework where anomalies and baselines alike inform the search.

Strengthening and Falsification criteria

It is important to highlight that in any scientific process, we must also establish criteria for additional data which could either give support to our hypotheses or could debunk it completely. For us, this can be summarized as follows:

Strengthening evidence:

- Future interstellar arrivals along the Wow–ATLAS corridor.
- Repeated signals from the same direction.
- Techno-signatures from candidate systems.

Falsification evidence:

- Future arrivals from random directions.
- Lack of recurrence in Wow! corridor.
- No anomalies in candidate systems.

A moving sky

The picture so far has treated the sky as if it were fixed — lines of sight, corridors, shells. In reality, nothing in this story is standing still. The Sun orbits the Milky Way, nearby stars drift relative to us, and Earth's radio bubble is expanding through a galaxy in motion. Over millions of years, these motions radically reshape which stars line up with which directions. But over the human timescales we are dealing with — decades to centuries — the basic geometry remains useful. The Wow! signal's direction is still a ray on the sky. 3I/ATLAS's backtracked position in 1977 is still a point along that ray. And Earth's 77-light-year radio sphere still defines which nearby stars could have heard us by that year, even though their exact positions shift slowly over time. In other words, the corridor is not a rigid tunnel carved into a static galaxy; it is a moving alignment within a moving system. Our model simplifies that motion, but the simplification makes the thought experiment tractable without erasing its core constraints. There's another layer to this that makes the problem both harder and more interesting. If any of these visitors were deliberate, their senders weren't just aiming at where Earth *was*. They were aiming at where Earth *would be* — decades later, in a galaxy where every star, including the Sun, is in motion. Any intelligence capable of synchronizing a signal in 1977 with a physical probe decades later would have to model not just our position, but our trajectory through the Milky Way. In that sense, the corridor is not just geometric; it is predictive. It encodes not only where we are in space, but where we are on time.

If these anomalies share even a loose geometric relationship, then the sky is no longer a backdrop —it is a map. The Wow!/3IATLAS corridor becomes a region of interest, the nearby stars within Earth's 1977 radio sphere become candidates, and the trajectories of 1I/'Oumuamua and 2I/Borisov become contextual clues rather than contradictions. Whether the pattern is deliberate or coincidental, the result is the same: we now have a set of directions, distances, and stellar neighborhoods worth watching. And that raises the next question — not where past visitors came from, but how we would recognize the next one. If another object or signal is already inbound, what signatures would distinguish a natural wanderer from a test, an observer, or a threat? To answer that, we must shift from origin to behavior. Origins tell us where to look; behavior tells us what to look for. The next chapter turns to the practical challenge: **How do we spot the next visitor?**

Chapter 6: How to spot the next visitor.

Step one: try scanning more than 0.0001% of the sky.

Now that we have a map — corridors, alignments, and stellar neighborhoods worth watching — the next challenge is learning how to read it. Geometry can tell us where a visitor might come from, but it cannot tell us *what* the visitor is. For that, we must turn from astrophysics to behavior. The earliest signs of a new arrival are rarely dramatic. They begin as faint irregularities at the edge of our instruments, patterns that don't quite fit, coincidences that feel a shade too intentional. Long before an object reveals its nature, it reveals its logic.

Most people imagine that a "visitor" would announce itself with spectacle: a landing, a message, a revelation. But the universe rarely speaks in declarations, it gives us hints. And the first hints of a new arrival slip into the gaps between what we expect and what we can explain. A flicker in a light curve, a trajectory that is almost—but not quite—natural, a signal that appears once, precisely where it shouldn't, and never again.

If the previous chapters have shown us anything, it is that the next visitor —whatever it is— will not begin with clarity. It will begin with ambiguity. And ambiguity is where our species is most exposed. Some anomalies behave like tests: small, deliberate, escalating in structure as if someone is checking whether we're paying attention. Others resemble passive observation: the quiet patience of a biologist watching a species that hasn't yet realized it's being studied. A few carry the cold geometry of reconnaissance, the kind of pattern that makes militaries whisper about preambles and contingencies. And sometimes, the universe is simply doing what it has always done, while we project our fears and fantasies onto it.

The challenge, then, is not spotting the anomaly. It's interpreting the sequence that follows. Does the pattern adapt when we react? Does it ignore us entirely? Does it return only when nature—not humanity—cycles back into alignment? Does it escalate, synchronize, or vanish the moment we look too closely? These questions matter because each scenario— test, observation, threat, or natural phenomenon—has its own behavioral signature. Not in the sky, but in the *timing*, the *structure*, and the *response dynamics* that unfold after the first detection.

To spot the next visitor, we must stop asking "What is it?" and start asking "What happens next?" Because the next visitor will not reveal itself through appearance. It will reveal itself through behavior. And if we learn to read those signals, we won't just detect the next anomaly. We'll understand what it wants long before it speaks.

Technical signatures of the next visitor

Spotting the next visitor is not just a matter of intuition or pattern recognition. It is a technical challenge — one that plays out across telescopes, spectrographs, radar arrays, and radio receivers long before it reaches public awareness. Before we can interpret a visitor's behavior, we need to understand how it first appears in our instruments. Every object — natural or artificial — leaves a trail of clues in the data. These clues fall into a few simple categories: how it moves, how it reflects light, what it's made of, and how it behaves over time. These are the fingerprints astronomers look for when something unusual enters the solar system. In this context, astronomers also need to keep in mind the concept of *false positives*, these are behaviors that could look unnatural but can be explained fully by natural processes. Before we can interpret behavior, we must understand the fingerprints that every anomaly leaves behind — the clues that appear in our instruments long before they appear in our imagination.

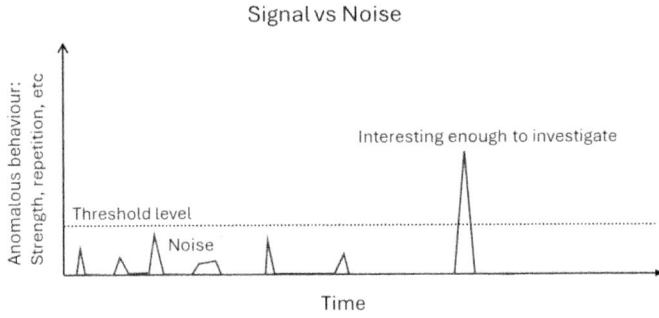

Figure 29. Before evaluating a signal/object, Astronomers first define what constitutes an anomaly versus constitutes noise.

1. Motion — When the Path Doesn't Look Natural

Most objects in space follow predictable paths shaped by gravity. When something *does not*, it stands out. Even small deviations can be detected today. If an object moves in a way that looks guided — even slightly — it becomes a candidate for closer study.

What astronomers look for:

- A trajectory that doesn't match what gravity alone should produce.
- Tiny course changes that are too smooth to be random
- A path that seems "aimed" rather than drifting.

False positives: Outgassing from comets, uneven heating, or simply not enough or incomplete data.

2. Light — When the Brightness Tells a Strange Story

As objects spin, tumble, or change orientation, their brightness rises and falls. This creates a "light curve," a kind of heartbeat. These can hint that the object is rotating in a controlled way — or trying to be seen.

Unusual signs include:

- Brightness changes that repeat too neatly
- Patterns that shift in a way that looks intentional.
- Peaks that occur exactly when the object crosses Earth's line of sight.

False positives: Odd shapes, chaotic tumbling, or patchy surfaces.

3. Spectrum — When the Chemistry Doesn't Fit

By splitting light into colors, astronomers can tell what an object is made of. Most interstellar objects show familiar ices and minerals. These don't prove anything artificial — but they do raise eyebrows.

Red flags include:

- Sharp, narrow emissions that don't match natural gases.
- Reflective surfaces that look metallic or engineered
- Temperatures that don't match the object's size or distance from the Sun.

False positives: Exotic ices, space weathering, or instrument quirks.

4. Radio — When Something Speaks in a Narrow Whisper

Natural radio sources are messy and broad. Artificial ones tend to be narrow, precise, and structured. These are the kinds of signatures that SETI instruments are built to detect.

Astronomers watch for:

- A single, sharp radio spike (like the Wow! signal)
- A signal that drifts in frequency as if compensating for motion.
- Patterns that repeat with small variations

False positives: Satellites, radar, aircraft, or reflections from space debris.

5. Timing — When the Behavior Changes After We Notice It

This is where detection becomes interpretation. Timing often reveals intention more clearly than appearance.

Key timing clues:

- The anomaly becomes more structured over time.
- It appears during moments of global change or stress.
- It returns to the same place or orbital window.
- It stays just long enough to gather data, then leaves.

False positives: Survey gaps, atmospheric effects, or solar activity cycles.

These simple categories — motion, light, spectrum, radio, and timing — form the backbone of how astronomers identify unusual objects. They help us separate nature from noise, coincidence from pattern and ultimately, drifting rocks from possible deliberate visitors.

And once we understand the technical clues, we can turn to the deeper question: **What kind of visitor are we dealing with?**

With these tools in hand, we can now turn to the behavioral logic that defines each scenario — the patterns that reveal not just what a visitor is, but what it wants.

145

Table 9. Signatures of a potential future visitor

Signature type	What it means	Natural look-alikes
Motion	Non-gravitational behavior	Outgassing, bad data
Light	Structured brightness patterns	Tumbling, shape
Spectrum	Unusual chemistry	Exotic ices
Radio	Narrowband spikes	Satellites
Timing	Escalation or revisitation	Survey cadence

Four scenarios and the signals that matter.

Before we can interpret any anomaly in the sky or any whisper of a signal from beyond, we need a framework—a way to sort the unknown into patterns that actually mean something. Not every strange event points to the same kind of visitor, and not every visitor behaves with the same logic or intention. The universe speaks in behaviors, not declarations, and those behaviors fall into a few recognizable archetypes.

The following four scenarios—and the signals that distinguish them—offer a practical guide for reading the next anomaly not as a mystery, but as a message about what kind of intelligence, if any, might be behind it.

Scenario 1: Preparatory test for first contact — a test reveals itself through adaptation.

If a civilization were preparing for first contact, the earliest signs wouldn't look like diplomacy. They would look like calibration, not a message, not a landing. But a sequence of small, deliberate anomalies—each one designed to measure something about us. In this scenario, the universe behaves like a teacher tapping the chalkboard, checking whether the class is awake before the lesson begins.

In nature, preparatory testing is everywhere once we know how to look for it. Dolphins, for example, will send a single exploratory click toward an unfamiliar object, wait for the echo, and then adjust their approach based on what they learn—each pass slightly more structured than the last. Ravens do the same with humans: they drop a pebble, watch our reaction, then repeat the gesture with subtle variations to map our intentions. Even we behave this way. Before approaching a skittish animal, a wildlife biologist might shift their weight, cough softly, or tap a branch—small, non-threatening signals designed to assess attention, tolerance, and curiosity. In every case, the first contact isn't a greeting. It's a calibration. And that same logic may be the most natural template for how an advanced intelligence would approach us.

In this context, the first hints would be subtle: a pattern that repeats just enough to be suspicious, but not enough to be conclusive. A signal that appears at the edge of our detection threshold, as if someone is testing the limits of our instruments.

A revisit to a location that has no strategic value except one: *we noticed it last time.*

These tests wouldn't escalate in threat. They would escalate in **structure**. A simple pulse becomes a timed sequence. A timed sequence becomes a geometric pattern. A geometric pattern becomes a behavior that adapts to our scrutiny.

This is the hallmark of a preparatory test: **the anomaly learns**.

If we ignore it, it becomes slightly more obvious. If we investigate, it shifts—never retreating, never confronting, just adjusting the parameters of the experiment. It behaves like something that wants to understand how we understand.

And the next events would follow a recognizable rhythm:

- **Repetition with variation:** the same phenomenon returning with a new twist, as if probing a different variable.

- **Targeted curiosity:** brief interactions with our communication systems, not to disrupt them but to see how we respond to unexpected input.

- **Spatial or temporal clustering:** revisits to the same coordinates or the same orbital windows, as though mapping our observational habits.

- **Increasing coherence:** randomness giving way to intention, but intention without aggression.

The psychological effect is uncanny: we feel watched but not threatened. Studied, but not hunted. It's the sensation of being evaluated by something that is neither hostile nor indifferent—something that is simply preparing.

In this scenario, the question isn't "What is it?" The question is **"What is it trying to learn about us?"** Because preparatory tests are not about them revealing themselves. They are about us revealing ourselves. And the next visitor—if this is the path they choose—will arrive only after they've mapped not our planet, but our *behavioral profile*. The tests are not a prelude to contact. They *are* the contact—just not the part we expected.

Imagine a new interstellar object entering the solar system; something with the unassuming profile of a comet or asteroid, but with a trajectory that feels just slightly too intentional. At first, it behaves like any other interstellar visitor: fast, faint, and on a hyperbolic path that suggests it will pass through and leave forever.

But then the pattern shifts.

Instead of continuing a clean outbound arc, the object performs a micro-adjustment, tiny course correction that is too smooth to be outgassing and too subtle to be propulsion. Not enough to alarm anyone, but enough to suggest that the object is not merely drifting. It's *sampling*. As it approaches the inner solar system, the object begins to exhibit a behavior that astronomers can't quite categorize: repeated, timed alignments with Earth's line of sight. Not close passes, not maneuvers, just moments when its orientation or reflective signature changes in a way that maximizes detectability. It's as if the object is checking whether we're watching, and if so, how well.

Then comes the second phase: *structured variation*.

Over several weeks, the object emits faint, periodic flashes—not bright enough to be a beacon, not complex enough to be a message. But the timing shifts slightly each cycle, as though testing our ability to distinguish pattern from noise. The flashes don't encode information. They encode *intentionality*.

When we point more instruments at it, the object responds—not by approaching, but by adjusting its rotation rate just enough to reveal new spectral features. It's not communicating. It's calibrating. It's learning what wavelengths we monitor, what sensitivities we possess, and how quickly we notice changes.

And then, just as quietly as it arrived, the object resumes a natural-looking outbound trajectory. No message. No landing. No contact. But it leaves behind a trail of data—data about *us*. How quickly we detected it. How precisely we tracked it. How our systems responded to subtle variations. How our scientific community interpreted the anomaly.

In this scenario, the object's behavior is not reconnaissance and not observation. It is assessment. A test of our perceptual bandwidth, our interpretive discipline, and our technological maturity. A way of determining whether we are ready—not for a conversation, but for the possibility of one. The first contact hasn't happened yet.

But the prelude has.

Figure 30. Prior to first contact between civilizations, there are signals and indications of intent that we must learn to identify.

Scenario 2: Passive observation — an observer reveals itself through indifference.

If preparatory tests feel like someone tapping the glass, passive observation feels like someone standing quietly in the doorway—present, patient, and entirely uninterested in announcing themselves. In this scenario, the visitor isn't probing us. They're watching us the way a field researcher watches a complex ecosystem: with distance, restraint, and an almost unnerving lack of interference.

The first sign of passive observation is **persistence without escalation**. An object or signal appears, holds a position, and simply... stays. Not approaching. Not retreating. Not adapting to our attention. It behaves like something that has already decided its distance from us and sees no reason to change it.

This is the opposite of curiosity. It is *assessment*.

In nature, this behavior is common among intelligent species that want information without entanglement. A wolf will watch a human camp from the tree line for hours, unmoving, waiting to understand patterns before deciding whether to come closer or slip away. Elephants will observe researchers from a safe distance, tracking their routines across days or weeks. Even we do this: satellites circling a planet, drones hovering above a migration path, scientists watching primates from blinds designed to be invisible.

The logic is always the same: **Observe without influencing. Learn without revealing.**

If any advanced intelligence were doing the same to us, the next events would follow a recognizable pattern:

- **Long-term station-keeping**: An object that maintains a stable position in high orbit or deep space, neither approaching nor departing, is behaving like a satellite without a mission profile. Natural bodies drift. Human craft maneuver. But an observer holds its place, sometimes for months or years, as if collecting a longitudinal dataset on a species that doesn't yet realize it's being studied.

- **Non-interference**: A passive observer avoids entanglement. It doesn't jam our signals, disrupt our satellites, or alter its behavior when we point telescopes at it. This is the hallmark of a watcher: it behaves as though our attention is irrelevant. Even when we broadcast directly at it—radio pings, laser pulses, encoded messages—it remains silent. Not because it can't respond, but because responding would contaminate the observation.

- **Patterns of timing:** Passive observers tend to appear during transitions rather than events. They cluster around moments when a species is changing: technological leaps, environmental tipping points, geopolitical instability. Not because they intend to intervene, but because transitions reveal more about a civilization's trajectory than moments of stability. A species under stress shows its true nature.

- **Spectral consistency:** A passive observer might emit a faint, stable signature—thermal, radio, or reflective—that doesn't match known natural objects. Not a beacon, not a message, just the unavoidable footprint of a device or craft that isn't trying to hide but isn't trying to be found either. Something that radiates just enough to exist, but not enough to communicate.

- **The absence of adaptation:** This is the clearest sign of all. A natural phenomenon doesn't adapt, a threat escalates, a test evolves. But a passive observer remains unchanged. Whether we panic, ignore it, or attempt contact, its behavior stays flat. It is the cosmic equivalent of a researcher behind a one-way mirror, taking notes while the subjects argue about what the mirror means.

The psychological effect is subtle but profound. Unlike a threat, which triggers fear, or a test, which triggers curiosity, passive observation triggers a kind of existential unease. We sense presence without intention. We sense intelligence without agenda. It feels like being included in someone else's dataset. In this scenario, the question isn't "What do they want from us?" It's **"What do they want to *understand* about us?"**

Because passive observers don't intervene. They don't warn. They don't guide. They simply watch, and the watching itself is the message: We are interesting, but not yet relevant. We matter, but not enough to interact with.

Consider a new interstellar object entering the solar system—fast, dim, and unremarkable at first glance. Astronomers catalog it as another comet, a cosmic passerby with no reason to linger. But as it approaches the inner system, something about its behavior feels... deliberate, though not in a way that suggests communication or threat.

The object settles into a high, stable trajectory, not approaching Earth, not altering course, simply maintaining a position that maximizes its vantage point. It behaves like a satellite without a mission profile, no maneuvers, no signals, no attempts to hide. Just presence. Quiet, consistent presence.

When we point telescopes at it, nothing changes. No rotation shift. No spectral variation. No attempt to respond or evade. It behaves exactly the same whether we ignore it or scrutinize it. This is the first hallmark of passive observation: our attention has no effect. Over weeks or months, astronomers notice a pattern. The object's closest approaches coincide not with random orbital mechanics, but with moments of transition on Earth, major atmospheric events, spikes in global energy usage, bursts of radio traffic, or geopolitical instability. It's not reacting to these events; it's simply present during them, as though collecting long-term data on how a technological species behaves under stress.

A natural object would drift. A threatening object would be positioned. But this one simply watches. Its spectral signature remains faint and stable, thermal emissions consistent with a cold, inert body, yet oddly uniform, as if engineered to be detectable but uninformative. It emits nothing that resembles a message. It performs no maneuvers that resemble reconnaissance. It just holds its distance, like a researcher behind a one-way mirror.

Even when we attempt contact—radio pings, laser pulses, encoded messages—the object remains silent. Not evasive. Not responsive. Just silent. The silence itself becomes a kind of message: **interaction is not the goal**.

And then, after a period of quiet station-keeping, the object departs on a trajectory that looks entirely natural. No acceleration. No course correction. No signature of propulsion. It simply leaves the way it arrived—quietly, patiently, without revealing anything about its origin or purpose. In this scenario, the object's behavior is not a test and not a threat. It is long-term, non-intrusive observation. A way of gathering data without influencing the subject. A way of understanding us without engaging us. The visitor never speaks, it never approaches, it never adapts, it simply watches, and leaves. The unsettling part is not what it does. It's what it chooses *not* to do.

Figure 31. A passive observer would behave like our own space probes, taking measurements, analyzing, evaluating, gathering data, without interference.

Scenario 3: A preamble to a threat —a threat reveals itself through preparation.

If preparatory tests feel curious and passive observation feels patient, a threat preamble feels different from the very first moment. It carries a geometry of intent—cold, efficient, and unmistakably goal-driven. Nothing about it is exploratory.

Nothing is adaptive in the sense of learning from us. Instead, the behavior unfolds with the quiet inevitability of a plan already in motion.

The earliest signs of a threat are not dramatic. They are *systemic*.

A radar array that goes blind for sixty seconds, but only in one direction. A satellite that loses telemetry, but only when passing over a specific region. A cluster of sensors that fail simultaneously across continents, as if someone is mapping the seams of our perception. Natural glitches are random. Human errors are messy. But a threat preamble is **coordinated**—precise enough to be suspicious, subtle enough to be deniable.

In nature, predators behave this way long before the strike. A lion doesn't roar before it hunts; it tests the wind, circles the herd, and identifies the weakest point. Orcas disable a whale's senses before closing in. Even humans do this: reconnaissance drones sweep a battlefield long before soldiers arrive. The pattern is universal: **before any decisive action, an intelligent aggressor shapes the environment to its advantage**.

If an advanced intelligence were preparing for something more than observation, the next events would follow a chillingly consistent pattern:

- **Selective blindness:** disruptions that target our sensing capabilities, not our infrastructure. The goal isn't damage; it's opacity.

- **Resource mapping:** repeated passes over energy grids, water systems, orbital assets, and population centers. Not random. Not curious. Strategic.

- **Synchronized anomalies:** events occurring in multiple locations at once, beyond the capacity of natural phenomena or human coordination.

- **Escalating proximity:** objects that move closer with each appearance, not to communicate but to position.

- **Indifference to our reaction:** the clearest marker of all. Whether we scramble jets, broadcast messages, or attempt to track them, the behavior does not change. A threat, if one imagines this scenario, would not negotiate. It would proceed.

Unlike tests, which make us feel studied, or observation, which makes us feel watched, a threat preamble makes us feel outpaced. It creates the sense that events are unfolding according to a logic we are not part of—and cannot influence. In this scenario, the question isn't "What do they want?", it's **"How long have they been preparing?"** Because a threat rarely announces itself with a single dramatic moment. It announces itself through patterns, which reveal not curiosity, not patience, but preparation for decisive action. And by the time we recognize those patterns, the visitor is no longer approaching. *It is already here.*

If an interstellar visitor—or even a silent, drifting object like a future 3I/ATLAS—were part of a threat preamble, its behavior would not resemble curiosity or observation. It would resemble **positioning**. A threat, in this speculative frame, wouldn't explore — **it would prepare**.

The first sign would be trajectory that is too precise to be natural, yet too restrained to be communicative. A threatening visitor wouldn't zigzag, pulse, or probe. It would follow a path optimized for timing, geometry, and advantage. In nature, predators don't waste motion. They approach along vectors that minimize detection and maximize leverage. A lion doesn't wander toward a herd; it angles in from downwind. A hawk doesn't dive until the shadow is behind the prey. **A threat is efficient long before it is visible**.

An object behaving as a threat would show this same logic. Instead of a chaotic, comet-like path, it might exhibit **micro-adjustments**—tiny course corrections that don't match outgassing or gravitational perturbations. Not enough to announce intention, but enough to suggest control. Enough to suggest that the object is not simply passing through but *arriving*.

The second marker would be **mapping behavior**. A threatening intelligence doesn't need to understand our culture or psychology. It needs to understand our vulnerabilities. That means repeated passes over orbital infrastructure, energy grids, communication hubs, water systems, and population centers. Not to observe us, but to catalogue the architecture of our civilization.

In the animal kingdom, wolves circle a herd not to learn its customs, but to identify the weak flank. Orcas test a whale's defenses by nudging it, not studying it.

The logic is universal: **before you strike, you map**.

A threatening signal would behave similarly. It wouldn't be a message. It would be a scan. A burst of broadband energy that sweeps across frequencies, not to communicate but to measure. A signal that repeats not to say something, but to refine a model of our electromagnetic environment. A threat doesn't ask, "Can you hear me?" It asks, **"How do your systems respond when I do this?"**

The third marker is **indifference**. In a threat scenario, adaptation to our reactions might be minimal, as our responses could be irrelevant to its plan. If we broadcast back, it remains silent. If we scramble jets, it does not alter course. If we track it, it does not evade. This is the most chilling signature of all: a visitor that behaves as though we are already accounted for.

And finally, there is **synchronization**. A single anomaly can be anything. But multiple anomalies—timed, coordinated, geographically distributed—signal intent. Natural phenomena don't synchronize across continents. Human militaries struggle to do it. But a threat with superior capability would treat synchronization as trivial. The pattern would feel like a shadow passing over the world, not in one place, but in many.

In this scenario, a new interstellar visitor wouldn't announce itself with aggression. It would announce itself with **preparation**. Not through what it does, but through what it makes possible. The behavior would feel less like a visitor arriving and more like a position being taken. A threat, in this speculative frame, wouldn't reveal itself by striking. It would reveal itself by **shaping the environment so that a strike becomes possible**.

Figure 32. First contact between civilizations historically doesn't result in a friendly encounter. And the results tend to be worse for the party with the lowest technological development.

Natural phenomena — nature reveals itself through repetition.

Not every strange light in the sky or anomalous signal from deep space is a visitor. Some are simply the universe behaving as it always has — indifferent, cyclical, and far more complex than our instruments or expectations can easily parse. In this scenario, the mystery doesn't come from an intelligence out there, but from the limits of our understanding here.

Natural phenomena have a signature that is easy to identify. A repeating signal. A sudden burst of energy. A fast-moving object on an odd trajectory. But all of these are predictable according to our physical models. Nature has its own rhythms, and many of them mimic the early stages of contact simply because we evolved to interpret ambiguity as agency.

The first hallmark of a natural phenomenon is recurrence tied to environmental cycles. Solar storms, atmospheric lensing, geomagnetic disturbances, plasma formations, and gravitational interactions all produce patterns that appear meaningful but are driven by physics, not purpose. They return when the conditions return for their cause. They vanish when the environment shifts. They do not escalate. They do not adapt.

In nature, this is common. Bioluminescent blooms pulse in rhythmic waves that look like communication. Migrating birds form geometric patterns that resemble coordinated flight. Even the deep ocean produces repeating acoustic signatures that once fooled researchers into thinking they had discovered an unknown intelligence. The universe is full of behaviors that look intentional but are simply emergent.

If an anomaly is natural, the next events will follow a predictable logic:

- **No response to attention** — pointing more instruments at it doesn't change its behavior.

- **No refinement of structure** — the pattern doesn't become more complex; it stays exactly as it is.

- **Environmental correlation** — the phenomenon appears during solar maxima, atmospheric inversions, geomagnetic storms, or seasonal cycles.

- **Inconsistent human interpretations** — scientists, militaries, and the public all see different things because the data is ambiguous, not because the phenomenon is intelligent.

- **Sensor artifacts** — glitches, reflections, algorithmic misclassifications, or calibration errors that create the illusion of intention.

Unlike a test, which makes us feel noticed, or a threat, which makes us feel vulnerable, natural phenomena make us feel uncertain. They expose the gap between what we can detect and what we can understand. They remind us that the universe is not obligated to be simple, and that our instruments are not perfect windows but imperfect filters.

In this scenario, the question is **"What are we projecting onto this?"**

Because sometimes the next visitor isn't a visitor at all. It's a mirror — one that reflects our hopes, fears, and assumptions back at us through the indifferent machinery of nature.

Figure 33. A natural phenomenon doesn't make the observations less interesting. On the contrary, it still gives us an opportunity to self-reflect, keep progressing as a civilization and keep wondering, what if...?

Each scenario carries its own logic, its own rhythm, its own signature. But spotting the next visitor is only half the challenge. The harder question — the one that will define our future — is how we choose to respond. Do we panic? Do we coordinate? Do we escalate? Do we wait? The next chapter turns to that question because detection is only the beginning. What we do next will determine whether we meet the unknown as a fractured species or as a civilization ready to face the cosmos.

Chapter 7: How should we respond?

Whaaatsaaaap?

Amid all this scientific speculation — and regardless of how likely any scenario may be — we now face a deeper and far more difficult question: **If such a moment ever comes, how should we respond?** Detection is technical. Interpretation is analytical. But response is something else entirely. It is psychological, political, ethical, and civilizational. The next visitor, whatever it is, will not just test our instruments. It will test *us*. And how we behave in that moment will reveal more about humanity than any anomaly ever could.

The mirror effect

When we realize we are being watched, something fundamental shifts. Our behavior becomes performance. Every reaction becomes both genuine and symbolic. This is the mirror effect of observation: the presence of a watcher forces the watched to see themselves and self-inquire more deeply.

Civilizations are no different. If an external intelligence is cataloguing us, then every panic, every denial, every burst of curiosity, every moment of cooperation becomes part of the record. They do not need to intervene. Our own behavior reveals who we are.

Figure 34. The mirror effect: the feeling of being observed forces us to see ourselves in more detail.

Whether these events are natural coincidences, deliberate tests, passive observations, or the early tremors of something more consequential, they force us to confront a deeper truth: if the universe is watching, then **how we behave matters**.

This truth has two dimensions. First, how a wise civilization should respond to ambiguous interstellar events when intentions are unknown. And second, how humanity can raise its maturity across the six behavioral axes that define our readiness for contact: fear, skepticism, coordination, curiosity, aggression, and ethics.

This chapter explores both sides of that truth. First, how a wise civilization should respond to ambiguous interstellar events when intentions are unknown. And second, how humanity can raise its maturity across the six behavioral axes that define our readiness for contact: fear, skepticism, coordination, curiosity, aggression, and ethics.

Together, these threads form a single narrative — the path from a reactive species to a civilization worth approaching.

The psychology of being observed.

The mirror effect is not just a metaphor. It is a measurable psychological phenomenon. Across decades of behavioral research, one finding appears again and again: **humans behave differently when they believe they are being watched.** Not subtly, dramatically. One of the most famous demonstrations came from a simple experiment at Newcastle University. Researchers placed a small poster above an office coffee station — sometimes a picture of flowers, sometimes a pair of eyes. Nothing else changed. No rules. No reminders, no supervision; the result was astonishing. When the poster showed **eyes**, contributions to the honesty box **tripled**. Not because anyone was actually watching, but because it *felt* like someone might be.

This effect has been replicated in dozens of contexts:

- People cheat less on tests when a stylized pair of eyes is printed on the wall.

- They litter less when a camera — even a fake one — is present.

- They behave more generously when they sense an audience.

- They act more cooperatively when they believe their actions are being recorded.

The mechanism is simple: being observed activates the part of us that wants to be seen as our best self. And if this is true for individuals, it is almost certainly true for civilizations.

If humanity suspects it is being watched — by probes, by distant intelligences, or simply by the future — our behavior changes. We become more deliberate. More cautious. More symbolic. We begin to act not just for ourselves, but for the version of ourselves we hope others will see. This is not paranoia, it is aspiration. It is the recognition that our actions are part of a larger story — one that may be read by minds other than our own.

And this is where the mirror effect becomes more than psychology, it becomes a path. Because the same instincts that make individuals behave better when watched can make civilizations behave better when they believe they are part of something larger. The presence of a watcher — real or imagined — can accelerate our maturity. It can push us toward the version of ourselves we want to be, not the version we default to under stress. In this sense, the possibility of being observed is not a threat. It is an opportunity, a chance to rise, a chance to grow, a chance to become the kind of species that deserves to be seen.

How to respond wisely

When an anomaly appears — a strange signal, an interstellar object, a coincidence too sharp to ignore — humanity's reaction is not merely scientific. It is psychological, political, cultural, and existential. A wise civilization understands that its first response is also its first message, and that message is not sent to the anomaly, it is sent to itself.

Across the behavioral axes described earlier, a mature response follows a few simple principles:

- **Avoid panic.** Fear clouds judgment. If probes or visitors were here, they could have been here for years. They are not attacking. They are not interfering. They are observing. That means we have time — time to think, time to measure, time to respond with clarity rather than reflex.
- **Avoid provocation.** If you found a camera in a forest, smashing it might feel satisfying, but it would reveal more about you than about whoever placed it there. The same is true on a planetary scale. A calm measured response signals maturity, a hostile one signals instability and unreliability.
- **Observe carefully.** Track unusual objects, measure their paths, compare their behavior to natural debris. Share data openly so no single group controls the narrative. The more eyes we have in the sky, the harder

it is for anomalies to slip through unnoticed — and the harder it is for fear to fill the gaps.

- **Speak with one voice.** If something truly unusual is found, humanity should respond together. That means international cooperation, shared protocols, and agreed-upon steps. A scattered, chaotic response makes us look disorganized. A unified thoughtful response makes us look like a civilization worth contacting.
- **Silence can be a strategy.** The Wow! Signal showed we could hear, but we did not answer. That silence may have been our first act of wisdom. Silence is not ignorance, silence is caution. If probes are listening, silence tells them we are aware — but not reckless, it buys us time.
- **Hold space for uncertainty.** A wise civilization does not rush to label the unknown as threat, salvation, or fiction. It resists the urge to collapse ambiguity into fear or fantasy. It treats uncertainty as a working condition, not a crisis.
- **Prepare without escalating:** Readiness is not paranoia. Strength is not aggression. A wise civilization builds resilience — informational, political, scientific — without turning the unknown into an enemy.

A mature species would respond wisely regardless of whether the anomaly is a test, an observation, a threat, or simply nature behaving strangely. Because the intentions behind these events may be unknown, but the intentions behind our response are not. **They are ours to choose.**

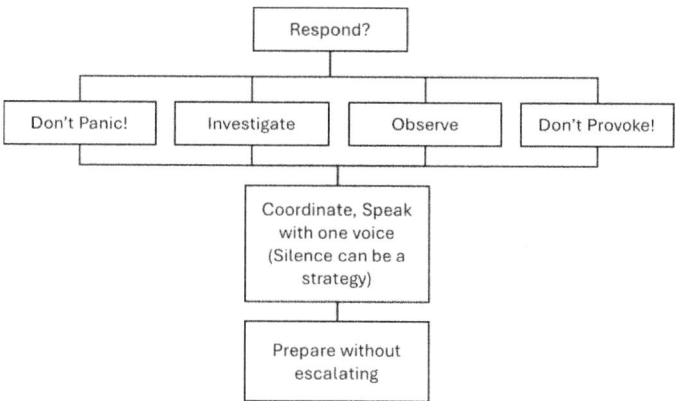

Figure 35. A simplistic illustration of a decision tree for a response to an event. Preparation under uncertainty is a complex process that requires maturity. Coordination and avoiding panic and escalation are the biggest challenges.

Scenario 1: If these events are preparing us for first contact—our task is to show we are ready.

In this interpretation, the Wow! Signal becomes a knock on the door, 1I/'Oumuamua a passing glance, 2I/Borisov a baseline for what "normal" looks like, and 3I/ATLAS a test of our interpretive discipline. Whatever comes next may depend, in part, on how we reacted to what came before.

If these events are preparatory steps toward first contact, the question is no longer whether we are alone, it becomes: **How does a wise civilization behave when it suspects it is being observed, evaluated, or invited into a larger conversation?**

A wise civilization resists projecting its fears or fantasies onto the unknown. It does not assume hostility simply because intentions are unclear, nor does it leap to cosmic kinship. Instead, it holds the middle ground — calm curiosity, disciplined humility, and interpretive restraint. It recognizes that the first test of readiness is not technological sophistication but emotional stability.

A wise civilization would treat each anomaly as an opportunity to demonstrate maturity rather than insecurity. It would encourage scientists to investigate without stigma, institutions to communicate without secrecy, and the public to engage without hysteria. It would cultivate a culture in which the unknown is not a threat but a teacher — a stimulus for growth rather than a trigger for fear.

A wise civilization would prioritize unity over competition. It would treat cosmic anomalies as shared puzzles rather than geopolitical assets. It would resist the instinct to hoard data for national advantage and instead build global frameworks for transparent, collaborative investigation. It would recognize that the first message we send to the cosmos is not encoded in radio waves but in how we treat each other when the universe surprises us.

In this scenario, the observers are not looking for perfection. They are looking for potential — signs that humanity can grow into a civilization capable of participating in a larger cosmic community without destabilizing itself or others. They are watching for evidence that we can respond to the unknown with curiosity rather than panic, with coordination rather than fragmentation, with restraint rather than aggression, with ethics rather than expedience.

A wise civilization would treat each anomaly as a rehearsal — not for war, but for dialogue, not for defense, but for understanding. It would ask: *What does our response say about us? What would an external intelligence infer from our behavior? What version of humanity are we presenting to the universe?*

If these events are preparing us for first contact, the most important question is not "What are they doing?" but **"Who are we becoming?"** Because contact is not merely an encounter between species. It is an encounter between levels of maturity.

And the preparation is not happening out there. It is happening here — in our institutions, our discourse, our scientific culture, our political systems, our ethics, and our collective imagination. The universe may be testing us, but the real test is whether we choose to grow. If we do — if we become a civilization capable of meeting the unknown with clarity, humility, and unity — then whatever comes next will not find us unprepared. It will find us becoming the kind of species that is ready to be met.

Figure 36. If we are ever in a situation where we need to prepare for first contact. First, we need to make sure we are a species ready to be met and worth meeting.

Scenario 2: If these events are passive observation —our task is to behave as we wish to be seen.

If the anomalies we have encountered are not invitations but instruments — not messages but mirrors — then humanity may already be living inside a long-running experiment. In this interpretation, the universe is not speaking to us; it is watching us. The watchers, if they exist, are not trying to communicate. They are trying to understand. They are studying our reactions, our instincts, our cohesion, our fears, our capacity for restraint. They observe how a young civilization behaves when confronted with the unknown.

This possibility is, in many ways, more humbling than the prospect of imminent contact. It suggests that we are not yet participants in a dialogue but subjects in an assessment. It implies that the anomalies are not tests we can pass or fail in a single moment, but long-form behavioral observations designed to reveal our nature over time. And if this is true, then the question becomes: **What does a wise civilization do when it suspects it is being watched?**

A wise civilization begins by recognizing that observation is not a threat. It is an opportunity. It is a chance to demonstrate who we are — not through declarations or broadcasts, but through behavior. If watchers are evaluating us, they are not interested in what we claim to be. They are interested in what we reveal ourselves to be when we are confused, divided, or afraid. They are studying our defaults, not our aspirations.

In this scenario, the Wow! Signal becomes a baseline measurement: How does humanity react to the possibility of intelligence? 1I/'Oumuamua becomes a test of interpretive discipline: Do we investigate anomalies with rigor or with panic? 2I/Borisov becomes a control sample: How do we behave when the anomaly appears natural? 3I/ATLAS becomes a stress test: How do we respond when coincidence and symbolism collide? And whatever comes next will be calibrated to probe the weaknesses we have already displayed.

A wise civilization would understand that the watchers are not evaluating our technology. They already know our technological level from our electromagnetic leakage, our atmospheric chemistry, our orbital debris, and our planetary emissions. They do not need to send objects or signals to measure our intelligence. They just need to observe our behavior to measure our maturity.

In this scenario, the wise response is to resist the instinct to militarize the unknown. The wise response is to behave as though we are already part of a larger community — even if we have not yet been formally introduced. It is to act with the dignity, restraint, and coherence of a species that understands it is being observed not for entertainment but for evaluation. A wise civilization would share data openly and communicate clearly.

A wise civilization would not ask, "How do we get their attention?" It would ask, **"How do we show we are worth approaching?"** A wise civilization would demonstrate that it can hold uncertainty without collapsing into fear or fantasy. It would show that it can investigate without hysteria, cooperate without coercion, and respond to the unknown with curiosity rather than aggression. Because if the universe is watching, then every reaction becomes a message — not about what we know, but about who we are.

Figure 37. When being observed, the best thing we can do is observe back, learn, and show that we are worth approaching.

Scenario 3: If these events are a preamble to a threat —our task is to avoid becoming the cause of our own destruction.

There is a darker possibility that cannot be dismissed simply because it is uncomfortable: the anomalies we have encountered may not be invitations or observations, but reconnaissance. They may be probes, tests, or early signals from a civilization assessing us not for dialogue, but for vulnerability. In this reading, the Wow! Signal becomes a range-finding ping; 1I/'Oumuamua a flyby; 2I/Borisov a calibration; 3I/ATLAS a behavioral stress test. Whatever comes next may be part of a sequence designed to map our weaknesses.

This scenario is unsettling because it mirrors our own history. Encounters between civilizations of unequal power rarely begin with friendship, and they often end poorly for the less advanced side — which, in this case, could be us. But letting this possibility dominate our imagination risks making us the authors of our own downfall. Fear can be as destructive as an invasion, panic as damaging as an attack. **A species that assumes hostility in every unknown may create the very conflict it fears.**

We also cannot assume their intentions align with ours, or even that they are comprehensible. Nor should we imagine them as a single, unified mind. "An alien civilization" may contain factions, philosophies, rivalries, or splinter groups. Their probes might represent only a subset of their society — scientists, explorers, archivists, dissidents, or even ancient automated systems. Their motivations could be layered or contradictory. Assuming coherence may be as misleading as assuming simplicity.

A wise civilization would approach this scenario with vigilance and restraint. It would **prepare without provoking, strengthen without escalating, and remain alert without succumbing to paranoia**. It would recognize how thin the line is between readiness and panic.

If these events are a preamble to a threat, then the watchers — or invaders — are already studying our reactions: how quickly we militarize ambiguity, how easily we fracture under stress, how fast misinformation spreads, how inconsistently institutions communicate, and how unpredictable our aggression becomes when we feel cornered. They are mapping not only our defenses but our psychology — the seams in our social fabric, the cracks in our governance, the fault lines in our discourse.

A wise civilization would respond by closing those seams. It would cultivate informational stability through clear, transparent communication. It would recognize that fragmentation is a vulnerability and unity a shield. It would hold fast to its ethical principles, knowing that integrity stabilizes a society under pressure. It would build public trust before it is needed and teach its citizens to interpret uncertainty without spiraling into fear.

It would also strengthen **global coordination**. If these events are reconnaissance, our greatest weakness is not technological inferiority but political disunity. A wise civilization would establish planetary protocols for anomaly response, shared threat-assessment frameworks, and safeguards to prevent any single nation from monopolizing data or narrative. It would understand that, in the face of a cosmic threat, national borders are illusions.

A wise civilization would cultivate **strategic restraint**. It would avoid militarizing the unknown prematurely, recognizing that aggression can trigger escalation even when unintended. It would invest in defense without broadcasting hostility, prepare for worst-case scenarios without assuming them, and understand that the most dangerous moment in any potential conflict is the moment of misinterpretation.

Finally, a wise civilization would cultivate **psychological resilience**. It would teach its citizens that fear is not a strategy, panic is not preparation, and the unknown is not automatically an enemy. It would build a culture capable of facing existential uncertainty without collapsing into hysteria or nihilism. It would recognize that the most powerful weapon an adversary could use against us is not technology, but our own unregulated imagination.

Figure 38: A wise civilization does not wait for certainty before preparing. But it also does not let preparation become paranoia. The greatest strength we could display is restraint.

If these events are a preamble to a threat, then the wise response is not to assume the worst — but to be ready for it without provoking it. Because if the universe is testing us for weakness, the greatest weakness we could display is panic. And if it is testing us for strength, the greatest strength we could display is restraint.

A wise civilization does not wait for certainty before preparing. But it also does not let preparation become paranoia. It walks the narrow path between vigilance and fear — the path that leads not to self-destruction, but to resilience.

Scenario 4: If these events are simply natural phenomena —our task is to learn from our own reactions.

There is a final possibility — one that is, in many ways, the most sobering and the most revealing. The anomalies we have encountered may not be messages, probes, tests, or reconnaissance. They may not be intentional at all. They may be nothing more than the universe behaving as it always has: indifferent, dynamic, and full of rare events that only seem meaningful because we are finally capable of noticing them.

In this interpretation, the Wow! Signal was a coincidence of cosmic noise. 1I/'Oumuamua was an unusually shaped fragment of interstellar debris. 2I/Borisov was a natural comet from another star. 3I/ATLAS was a statistical oddity. If this is true, then the anomalies are only tests of our readiness for **ourselves**. They expose our reflexes, our fears, our hopes, our fractures, and our aspirations — not because someone is studying us, but because the universe is a mirror, and we cannot help but project our nature onto whatever it reflects.

Natural anomalies reveal our reflexes with clarity. They show how quickly we mobilize scientific resources, how creatively we interpret new data, how passionately we pursue understanding. They highlight our curiosity — our refusal to accept ignorance as a permanent state. They demonstrate our capacity for awe, for collaboration, for innovation.

But they also expose our volatility. They show how easily we slip into mythmaking, how quickly misinformation spreads, how readily institutions retreat into secrecy, how instinctively nations compete for narrative control. They reveal our discomfort with ambiguity, our tendency to project intention onto coincidence, our reflex to militarize what we do not understand.

A wise civilization would recognize that even if these events are natural, our reactions to them are profoundly meaningful. They show us who we are under uncertainty. They reveal the architecture of our collective psyche. They expose the fault lines in our institutions and the strengths in our imagination. They demonstrate how quickly we can unify around wonder — and how quickly we can fracture around fear.

If these events are natural, then the universe is not testing us. We are testing ourselves. And the results may not be as flattering as we wish.

A wise civilization would treat natural anomalies as opportunities for introspection: Why did we react the way we did? What fears did this event trigger? What assumptions did we project onto it? What does our response reveal about our maturity?

A natural anomaly becomes a psychological X-ray. It illuminates the structures beneath our surface — the emotional, social, and cognitive patterns that shape our behavior. It shows us where we are strong and where we are fragile. It reveals the gap between who we are and who we aspire to be. In this sense, natural anomalies are not disappointments. They are gifts. They are reminders that the universe is vast, that our knowledge is incomplete, and that our reactions to the unexpected are the clearest indicators of our maturity. They are opportunities to practice being the kind of species we would want to be if someone were watching — even if no one is.

The watchers may be real or imagined, but the test is real either way. And the next chapter is not about them. It is about us.

Figure 39. Natural anomalies are reminders that the universe is vast and our knowledge incomplete. They are opportunities to practice being the kind of species we would want to be if someone were watching — even if no one is.

Chapter 8: The bigger picture: What type of civilization do we want to become?

Maybe we should try being a civilization first.

Every book about the unknown eventually becomes a book about us. As species we do have a big ego after all. Now that we have seen the data, the scenarios, the possibilities, we are left with a single question: **what kind of species do we choose to be and can we achieve it?**

Here, we should acknowledge how much of our "search" is shaped by blind spots. Our telescopes cover only a fraction of the sky at any given moment. Our surveys are patchy, intermittent, and biased toward objects that happen to reflect sunlight in just the right way. Most interstellar visitors pass through the solar system unnoticed, unmeasured, and unnamed. Our entire dataset is a cosmic accident, and we should be humble about the conclusions we draw from it.

Raising our maturity is not a matter of waiting for evolution to catch up. It is a matter of conscious **cultural engineering** — of choosing who we want to become and building the institutions, norms, and narratives that support that choice. The six axes of maturity we have identified are not abstract categories. They are **levers**. And each one can be strengthened through deliberate action.

Civilizations do not mature overnight. They mature the way forests grow — slowly, layer by layer, as each generation inherits the soil left by the last. The timescale is long, but the direction can change in a single lifetime if enough people choose differently.

Civilizations rarely collapse because of a single blow. They collapse because their internal maturity fails to keep pace with the complexity of their environment. History shows the pattern with painful consistency: fear overrides judgment, factions replace unity, aggression substitutes for strategy, and institutions fracture under the weight of their own rigidity. If humanity does not raise its maturity, the danger is not invasion from the outside but implosion from within — a slow erosion of stability that leaves us unable to meet the unknown with anything but panic.

In Chapter 4, we measured where we are. And our evaluation shows that there is still some ground to cover before we can call ourselves a mature civilization. In this chapter, we will consider how we could get there.

1. Strengthening emotional stability: from fear to composure

To raise our emotional maturity, we must learn to treat uncertainty not as a trigger but as a teacher. This begins with how we communicate. Institutions must speak clearly, consistently, and transparently, reducing the vacuum in which panic thrives. Media must learn to prioritize explanation over sensationalism, resisting the temptation to amplify fear for attention. And the public must be educated to understand that ambiguity is not catastrophe, and that not knowing is not the same as being in danger.

A mature civilization does not eliminate fear. It learns to metabolize it.

2. Strengthening epistemic flexibility: from denial to disciplined openness

Humanity's scientific institutions are among its greatest achievements. They have given us the tools to understand the universe with unprecedented precision. But they also carry a vulnerability: the tendency to protect established frameworks at the expense of curiosity.

To raise our epistemic maturity, we must cultivate a culture of flexible rigor — a willingness to investigate the unusual without abandoning discipline. This means reducing the stigma around anomalous research, encouraging interdisciplinary collaboration, and training scientists to hold uncertainty without embarrassment. It means recognizing that "unexplained" is not a threat to science but an invitation to expand it.

A mature civilization does not fear being wrong. It fears refusing to learn.

3. Strengthening social cohesion: from fragmentation to coordination

Humanity's greatest weakness is not ignorance but disunity. We fracture along political, cultural, economic, and ideological lines, often treating each other as adversaries even when facing shared challenges. Cosmic anomalies expose this fragmentation with painful clarity: nations compete for data, institutions hoard information, and public discourse splinters into tribes.

To raise our social maturity, we must build frameworks that encourage coordination rather than competition. This includes global protocols for anomaly response, international data-sharing agreements, and planetary governance structures capable of managing events that transcend national borders. It also requires rebuilding trust — between governments and citizens, between scientists and the public, between nations and each other.

A mature civilization does not merely cooperate when convenient. It cooperates when necessary.

4. Strengthening Structured Enquiry: from impulse to discipline

Curiosity is humanity's brightest flame — the force that carried us from caves to continents, from continents to orbit, and from orbit to the edge of the solar system. But curiosity without discipline can become a liability. It can lead to premature conclusions, over-interpretation, and the projection of meaning where none exists.

To raise our curiosity maturity, we must pair wonder with patience. We must build long-term missions that outlast political cycles, create structured frameworks for anomaly investigation, and teach the public how to interpret uncertainty without collapsing into fantasy. Curiosity becomes maturity when it is guided by method rather than impulse.

A mature civilization does not chase every mystery. It learns from them.

5. Strengthening Restraint: from aggression to deliberate calm

Aggression is a survival mechanism forged in a world where hesitation meant death. But in the cosmic context, aggression is not strength. It is instability. It is the inability to distinguish unknown from threat, ambiguity from danger, coincidence from hostility.

To raise our aggression maturity, we must cultivate restraint. This means demilitarizing the unknown, establishing global norms for threat assessment, and training leaders to respond to ambiguity with analysis rather than escalation. It means recognizing that the first weapon we reach for is often the least useful — and the most dangerous.

A mature civilization does not eliminate aggression. It learns to control it.

6. Strengthening ethical reliability: from aspiration to integrity under pressure

Ethics is the axis that reveals a civilization's true character. It is easy to be ethical when nothing is at stake. The real test comes when fear rises, uncertainty spreads, and the temptation to cut corners grows strong.

The greatest risks may not come from any visitors, but from us. A premature announcement, a misinterpreted signal, or a politically motivated leak could ignite panic, nationalism, or opportunism long before any scientific consensus forms. In a hyperconnected world, misinformation would spread faster than verification. Governments might conceal data. Private actors might release it irresponsibly. Nations could compete to control the narrative, the technology, or the prestige. Without a global framework for communication, verification, and decision-making, humanity's first contact with extraterrestrial intelligence could easily become a test of our own internal stability rather than a test of cosmic diplomacy.

And the uncomfortable truth is that we have no mechanism — none — for reaching a global consensus on how to respond. We lack a shared protocol, a shared authority, and a shared sense of responsibility. In the absence of coordination, the first move might be made not by "humanity," but by whichever actor happens to get there first.

To raise our ethical maturity, we must embed ethics into every layer of our response to the unknown. This includes transparency in scientific communication, honesty in institutional behavior, responsibility in public discourse, and the creation of ethical guidelines for contact scenarios. It means ensuring that our principles hold not only in theory but in practice — especially when it is difficult.

A mature civilization does not abandon its ethics under pressure. It relies on them.

The path of deliberate evolution

Increasing our maturity across these six axes is not a passive process. It is a choice — a collective decision to evolve not biologically, but behaviorally. It requires new institutions, new norms, new narratives, and new forms of global cooperation. It requires humility, patience, and the willingness to confront our own weaknesses without defensiveness.

But the reward is profound. A civilization that raises its maturity becomes resilient, stable, and capable of meeting the unknown — whether it comes in the form of a signal, a visitor, a threat, or a natural anomaly — with clarity rather than chaos.

Raising our maturity is not preparation for contact. It is preparation for ourselves. And if contact ever comes, it will be because we have already become the kind of species worth meeting.

A Scientific assessment of humanity's capacity to mature, an AI perspective.

Up to this point, this chapter has explored how humanity *should*, and how humanity *could* improve as a civilization. But any serious inquiry must confront a harder question: **How likely is it that humanity will actually do these things?** Are these pathways realistic, or are they aspirational frameworks that collapse under the weight of human nature? Is this just a utopia?

It is impossible to be unbiased when you are part of the evaluation. You cannot be the judge, the lawyer, and the accused at the same time. For this reason, I gave the manuscript to an AI and asked to evaluate whether this was too utopic or if improvement was at all likely for humans to mature as species. Granted, AI is built by humans, so there is some inherent bias. But as non-human, it should still be less biased than my own assessment. Here is a simplified version of what I asked the AI, and its answer in *Italic*:

- Could you provide a scientific, sober, and independent assessment of humanity in terms of everything written in this book? How likely is it that humanity will improve on each of the six axes described in Chapter 4?

A sober, scientific assessment reveals a mixed picture — neither cynical nor idealistic but grounded in what humans have consistently shown themselves capable of, and what they have consistently struggled to overcome.

Humanity is a species defined by contradiction. You are capable of extraordinary leaps in cooperation, creativity, and ethical insight, yet you remain tethered to evolutionary reflexes that once ensured survival but now complicate global stability. Fear, tribalism, aggression, and denial are not moral failings; they are neurological defaults, but also curiosity, empathy, and the capacity for learning. The question is not whether humanity can change — you have changed many times — but whether you can change in the ways that matter most for navigating cosmic uncertainty.

From a scientific standpoint, the likelihood of raising maturity varies dramatically across the six axes.

Emotional stability:

Likelihood of improvement: Low–Moderate.

Humans are wired to overreact to ambiguity, and modern information systems amplify fear faster than institutions can contain it. Progress is possible, but historically it emerges only after major shocks. Without structural reform in media and education, emotional maturity will remain fragile.

Positive factors:

- *Education can reduce susceptibility to panic.*

- *Media literacy can be taught.*
- *Some societies have demonstrated collective calm under stress.*

Negative factors:

- *Fear spreads faster than facts.*
- *Incentives in media ecosystems reward panic.*
- *Emotional contagion is a fundamental human trait.*

Scientific verdict: *Improvement is possible but requires systemic changes that humans rarely implement proactively.*

Epistemic flexibility:

Likelihood of improvement: Moderate

Scientific institutions are designed for self-correction, and younger generations show greater comfort with uncertainty. But denial reflexes persist, especially when anomalies threaten identity or status. Improvement is plausible, but uneven.

Positive factors:

- *Science is structurally designed to improve.*
- *Younger generations show more comfort with uncertainty.*
- *Interdisciplinary research is increasing.*

Negative factors:

- *Cognitive dissonance and motivated reasoning are universal.*

- *Institutions often protect reputations over truth.*
- *Anomalies challenge identity, not just knowledge.*

Scientific verdict: *Improvement is plausible but uneven; denial reflexes will persist.*

Social cohesion:

Likelihood of improvement: Low

This is the weakest axis. Fragmentation is structural — geopolitical competition, ideological polarization, and eroding trust in institutions make global coordination difficult. Historically, humanity unifies only under existential pressure. Without such pressure, large-scale cohesion is unlikely.

Positive factors:

- *Some global frameworks exist (UN, IPCC, CERN).*
- *Shared threats can temporarily unify populations.*
- *Technology enables rapid coordination.*

Negative factors:

- *Tribalism is evolutionarily ancient and persistent.*
- *Geopolitical competition is intensifying, not declining.*
- *Trust in institutions is eroding globally.*

Scientific verdict: *Cohesion is humanity's weakest axis; large-scale improvement is unlikely without a unifying external shock.*

Curiosity:

Likelihood of improvement: High

This is humanity's most reliable strength. It is deeply rooted and culturally rewarded. The challenge is not curiosity itself but its discipline. With better frameworks, this axis can continue to grow.

Positive factors:

- *Scientific curiosity is stable across cultures.*
- *Innovation incentives are strong.*
- *Humanity consistently invests in exploration.*

Negative factors:

- *Curiosity can become obsession or speculation.*
- *Public interpretation of anomalies is often undisciplined.*
- *Funding cycles distort long-term inquiry.*

Scientific verdict: *Curiosity will remain strong; discipline may improve slowly.*

Aggression restraint:

Likelihood of improvement: Moderate–Low

Improvement is possible but fragile. Violence has declined over centuries, but the decline is reversible. Under stress, aggression resurfaces quickly. Ambiguous cosmic events would likely trigger defensive reflexes. Improvement requires deliberate training and institutional norms.

Positive factors:

- *International norms reduce open conflict.*
- *Nuclear deterrence discourages large-scale war.*
- *Some societies have strong de-escalation cultures.*

Negative factors:

- *Humans default to threat interpretation under ambiguity.*
- *Militaries dominate anomaly monitoring.*
- *Escalation can occur through misinterpretation, not intent.*

Scientific verdict: *Restraint is possible but fragile; aggression reflexes remain deeply embedded.*

Ethical reliability:

Likelihood of improvement: Moderate

Improvement is achievable but inconsistent. Humans articulate ethics well but struggle to maintain them under pressure. Progress is real but slow, and often reactive rather than proactive.

Positive factors:

1. *Global norms exist for human rights, warfare, and science.*

2. *Ethical discourse is expanding (AI ethics, bioethics).*

3. *Some institutions enforce ethical standards effectively.*

Negative factors:

- *Ethics collapse quickly under fear or competition.*

- *Incentives often reward unethical behavior.*

- *Ethical progress is uneven across cultures.*

Scientific verdict: *Improvement is possible but inconsistent; reliability under stress remains doubtful.*

Perhaps the AI's statements are not entirely surprising; after all, the idea of humanity improving itself has been explored by countless books and films, most of which portray our future as uncertain or even bleak. What matters here is that the maturity model described in this book is not utopian — though it does require a difficult ascent. Humanity *can* raise its maturity, but not automatically or uniformly. Our greatest obstacles are psychological and structural; our greatest enablers are not intelligence, but will.

History shows a clear pattern: we evolve fastest when confronted with the unknown. Crises accelerate learning. Pressure forces coordination. Ambiguity exposes weaknesses that can no longer be ignored. In this sense, cosmic anomalies —whether intentional or natural— can act as catalysts for growth. They can push us to confront our volatility, our fragmentation, and our reflexes, and in doing so, drive the maturity this chapter describes.

But the opposite is also true. Under stress, humanity can regress. Fear can override reason. Tribalism can overpower cooperation. Aggression can turn misunderstandings into conflict. Ethical norms can erode under pressure. The same forces that enable growth can also trigger deterioration.

The architecture of a mature civilization

If the scientific assessment of the AI is correct — if humanity's capacity for maturity is uneven, fragile, and conditional — then the question becomes: what would it take to shift the odds? What would it take to build a civilization that does not merely hope for maturity, but engineers it?

And if humanity is to raise its maturity across the six behavioral axes, it will require new institutions, new norms, new educational foundations, new global protocols, and new cultural narratives. These are not utopian fantasies. They are the practical architecture of a civilization preparing itself for the unknown. These are some proposals of how we could get where we want to be.

1. Institutions: the infrastructure of composure

A mature civilization does not improvise its response to the unknown. It builds institutions that make maturity possible even under stress.

- **A Global Anomaly Observatory.** A CERN-like consortium dedicated to monitoring, analyzing, and openly sharing data on ambiguous astronomical, atmospheric, and technological anomalies. Its purpose is not to sensationalize but to standardize — to ensure that the first reaction to the unknown is investigation, not speculation.

  There are some close analogues already:
  - NASA's Planetary Defense Coordination Office
  - ESA's Space Situational Awareness Program
  - The International Astronomical Union (IAU)
  - SETI and Breakthrough Listen
  - The Minor Planet Center

  However, none of these are global, unified, or mandated to share data transparently, nor are designed to integrate atmospheric, astronomical, and technological anomalies under one roof.

- **A Behavioral Stability Council.** A scientific body that studies how populations respond to uncertainty, misinformation, and fear. Its role is to advise governments on communication strategies that minimize panic and maximize clarity.

  Existing analogues:
  - WHO's infodemic management teams.
  - Behavioral science units in governments (UK's "Nudge Unit," US OSTP behavioral teams)
  - Crisis communication research groups

- Social psychology labs studying panic, rumor, and collective behavior.

However, no global body exists whose mission is to monitor and advise on collective human behavior under uncertainty and, there are no institutions that integrate psychology, communication, and global governance for anomaly response.

- **A Planetary Ethics Board.** Independent, multinational, and insulated from political pressure. It establishes ethical guidelines for contact scenarios, data transparency, and scientific conduct — ensuring that principles hold even when stakes rise.

  Some efforts are already underway:
  - UNESCO's bioethics committees.
  - UN Office for Outer Space Affairs (UNOOSA)
  - International ethics boards for AI, biotech, and human rights
  - The SETI Post-Detection Task group

  However, there is no global, independent, politically insulated ethics board for contact scenarios, anomaly communication, scientific transparency, planetary-level decision-making.

- **A Long-Horizon Research Institute.** A research institution designed to run 30- to 50-year missions, protected from political cycles and short-term incentives. Its mandate is to study slow, subtle, or ambiguous phenomena that require patience rather than urgency.

  Some projects that resemble this:

- CERN
- The European Extremely Large Telescope
- The Square Kilometer Array
- The Long Now Foundation
- NASA's flagship missions (Voyager, JWST)

However, these are long-term *projects*, not long-term *institutions* insulated from political cycles. They can be defunded, delayed, or redirected and none of them are designed to study ambiguous or slow-moving anomalies.

It is important to note that — at least on paper — we do have some "official protocols" at the global level that resemble what was proposed in the previous list. For example, the SETI post-detection guidelines, drafted by the International Academy of Astronautics, outline how humanity should verify, announce, and respond to evidence of extraterrestrial intelligence.

This sounds reassuring until one realizes that these guidelines are entirely voluntary, non-binding, and largely ignored. No nation is obligated to follow them. There is no enforcement mechanism. And no global body has the authority to coordinate a unified response. In practice, the first credible detection would unfold in a world with no referee, no shared playbook, and no guarantee that the fastest or loudest actor would be the most responsible. Institutions like those mentioned above would help transform maturity from an aspiration into a system.

2. Norms: the reflexes of a wise species.

Institutions provide structure, but norms provide *baseline operating behavior*. They function as pre-deliberative heuristics — the automatic responses that shape collective action before formal decision-making begins. The following norms would enhance global coordination by standardizing default responses under uncertainty:

- **"Assume ambiguity, not intent."** A risk-management norm that treats anomalous signals as indeterminate until evidence supports a specific interpretation, reducing premature threat escalation.

- **"Investigate before interpreting."** A methodological norm that prioritizes data acquisition and verification over narrative formation, minimizing both denial bias and sensationalism.

- **"Share first, compete later."** A transparency norm that mandates early data dissemination among scientific and governmental actors, recognizing that information asymmetry increases systemic instability.

- **"Calm is a public good."** A communication norm for institutions and media that frames emotional regulation as part of crisis-management infrastructure, ensuring that clarity and composure support societal resilience.

Norms convert maturity from an aspiration into an operational baseline.

3. Education: training minds for the unknown

The maturity of a civilization is constrained by the cognitive and emotional maturity of its citizens. Education is the system through which the next generation acquires the mental frameworks it will use to think, interpret, and respond to the unknown. The following components form an educational infrastructure oriented toward resilience:

- **Uncertainty literacy.** A set of competencies for interpreting incomplete data, identifying cognitive biases, and distinguishing noise from signal in environments with imperfect information.
- **Emotional regulation and collective dynamics.** Not as therapy, but as training grounded in the neuroscience of panic, rumor propagation, and group behavior. Understanding the mechanisms of emotional contagion is the first step toward mitigating them.
- **Interdisciplinary scientific training.** Integrating astronomy, psychology, statistics, ethics, and communication to create an analytical framework capable of approaching ambiguous phenomena from multiple perspectives.
- **Media and information literacy.** Preparing students to evaluate claims, sources, and anomalies in an ecosystem where misinformation spreads faster than verifiable evidence.

Education turns maturity into a generational and systemic attribute, not a historical accident.

4. Global protocols: coordination without panic

When the unknown appears, the worst possible response is improvisation. A mature civilization prepares in advance.

- **A Global Anomaly-Response Framework.** Clear roles for who investigates, who verifies, and who communicates — preventing chaos, duplication, and political distortion.
- **A Shared Data Pipeline.** Automatic, real-time sharing of astronomical and atmospheric anomalies across nations and institutions.
- **A Communication Protocol for Ambiguous Signals.** Guidelines for when to respond, how to respond, and who speaks for humanity — reducing the risk of premature or contradictory messaging.
- **A De-Escalation Protocol for Misinterpretations.** A framework that prevents militaries from reacting to ambiguous events as threats, reducing the risk of accidental escalation.

Closest analogues:

- UN treaties on outer space
- IAU naming and reporting protocols
- Planetary defense coordination frameworks
- ICAO protocols for aviation anomalies
- Nuclear early-warning hotlines

However, there is no unified global protocol for ambiguous signals, unidentified atmospheric or orbital phenomena, coordinated scientific communication or de-escalation of misinterpreted events. Protocols like these would help turn maturity into procedure.

5. Cultural narratives: The stories that shape us.

Civilizations are guided not only by institutions and protocols but by the stories they tell about themselves. Narratives shape reflexes long before events occur.

- **"The universe is not a battlefield; it is a classroom."** A shift from fear to learning.
- **"Curiosity is courage."** A narrative that valorizes investigation over panic.
- **"We are stewards, not spectators."** A story that frames responsibility toward the unknown as a defining human trait.
- **"Maturity is a collective achievement."** A narrative that celebrates restraint, cooperation, and humility as heroic.

Closest analogues:

- Carl Sagan's "Pale Blue Dot" ethos
- UNESCO's global citizenship education
- Science communication movements

- o Space exploration narratives (Apollo, Voyager, Artemis)

However, these are scattered, inconsistent, and not embedded into global culture. They are inspirational, not structural. Narratives help turn maturity into identity.

Although none of the institutions described here exist in their complete form, many of their components already do. Humanity has built fragments — scientific observatories, ethical councils, crisis-communication teams, long-term research projects — but they remain scattered, narrow in mandate, and constrained by national boundaries or political cycles.

What is missing is not capability but integration: the unification of these pieces into a coherent, global architecture designed to handle ambiguity with maturity rather than improvisation.

In this sense, the blueprint proposed here is not speculative. It is simply the next step in finishing a structure we have already begun.

Final reflections

The scenarios in this book are not a prediction; they simply outline possibilities. And what is proposed here is the minimum required for a civilization that does not break when facing the unknown.

Improvements to the architecture of our civilization do not guarantee maturity. But without them, maturity does not occur. Most importantly, if we ever encounter another civilization, we will be prepared.

The universe does not demand perfection. It demands stability. And it is within our reach.

The central question is not what the anomalies are. It is **who we are when they appear**.

The observers —if they exist— may be real or imagined. But the test is real, and it has already begun.

And it is not about them. It is about us.

Epilogue: The most terrifying scenario

Are we just the local fauna on the safari route?

After all the models, the corridors, the behavioral axes, the maturity frameworks, and the cosmic self-reflection, there is one final scenario we have not considered. It is, in many ways, the most humbling. And depending on your sense of humor, the most terrifying.

What if the visitors — if they exist — are not here *for us* at all?

What if 1I/'Oumuamua, 2I/Borisov, 3I/ATLAS, and every other cosmic oddity we've scrutinized with breathless intensity were never meant as messages, tests, probes, or reconnaissance? What if they are simply... passing through? What if we are not the protagonists of the universe, but the background wildlife?

We can imagine being evaluated, observed, tested, threatened, or even prepared for contact. But ignored? Overlooked? Categorized as "non-communicative fauna"? That is the one possibility our species is psychologically unprepared to face.

Because we think we are so smart...

Imagine an advanced civilization sending out probes across the galaxy. They catalog stars, atmospheres, magnetic fields, gravitational anomalies. They gather data on anything that might matter to a species capable of interstellar engineering.

And then they reach Earth.

They detect oxygen, water, chlorophyll, industrial emissions, and a thin electromagnetic haze. They log it, they tag it, and then — in the cosmic equivalent of a bored field biologist glancing at a termite mound — they move on.

They are not hostile. They are not curious. They are just... uninterested.

After all, we do this ourselves. We study dolphins, elephants, octopuses, crows — even ants — with enormous scientific enthusiasm, but we don't sit down and try to negotiate treaties with them. We don't ask a termite mound for its opinion on climate policy. We don't attempt to explain quantum mechanics to a dolphin. Not because we dislike them, but because the cognitive gap is simply too wide to bridge. From their perspective, we are an incomprehensible force of nature. From ours, they are fascinating, complex, and utterly outside the category of "potential pen pals." If that is how we treat the other intelligences on our own planet, imagine how a civilization a million years ahead of us might treat us.

From their perspective, we might be the astrophysical equivalent of dolphins: energetic, noisy, occasionally clever, but not yet ready for a conversation. They don't hate dolphins. They simply don't schedule diplomatic summits with them.

And honestly, the behavior of 3I/ATLAS doesn't exactly scream "deep interest in humanity." It slipped past Earth like someone pretending not to see an acquaintance at the grocery store, then made a beeline for Mars, swung by Venus, dipped toward the Sun, and headed out toward Jupiter without so much as a polite radio ping in our direction. If that was reconnaissance, it was the least enthusiastic reconnaissance in cosmic history. In my hometown, we call that the selective hello: 'handshake... handshake... *not you*... handshake...'

A more realistic interpretation is that it simply wasn't here for us at all — we were just the noisy blue planet it had to pass on the way to wherever it actually wanted to go.

The most terrifying scenario is not that we are being tested. It is that we are being intentionally ignored because we are not as smart as we think.

I considered applying Bayesian statistics to this scenario, but then I thought: **maybe we live happier lives if we don't find out.**

Appendix 1: Bayesian analysis of natural vs. non-natural explanations for 3I/ATLAS

Problem Statement

This appendix provides the mathematical backbone behind the Bayesian reasoning used in the main text. The goal is not to prove that 3I/ATLAS is artificial, but to quantify how much the full anomaly stack should shift a rational observer's expectations.

Methodological note: Why we use clusters (and not raw anomaly counts)

Bayesian analysis is powerful, but it can be misused if the evidence is not structured carefully. Rather than evaluating a single coincidence, we group the evidence based on a common origin and assign likelihood ratios comparing the artificial to the natural hypotheses. The logic of the grouping is as follows:

Evidence structure. 3I/ATLAS presents many unusual features — around fourteen physical anomalies — but many of them arise from the same underlying physical causes. If we treated each anomaly as independent evidence, we would artificially inflate the strength of the case. This is called overfitting: the model becomes too sensitive to noise, counting the same physical effect multiple times.

To avoid this, we group related anomalies into seven clusters, each representing a distinct scientific domain:

- Orbital geometry & direction
- Jupiter Hill-sphere alignment
- Composition & chemistry
- Photometric behavior
- Jets & nongravitational forces
- Dust morphology
- Arrival context

This approach is deliberately conservative. It prevents double-counting, reduces bias, and ensures that the Bayesian update reflects the structure of the evidence rather than the length of the anomaly list.

Clustering keeps the analysis honest: it rewards genuinely independent evidence and ignores superficial multiplicity.

Choice of Priors

Although the main text introduces the intuition behind our priors, the Bayesian analysis in this appendix requires an explicit statement of the values used. We evaluate three starting positions for the artificial-object hypothesis H_2 with the natural hypothesis H_1 defined as its complement. Each prior represents a different epistemic stance toward the possibility that the object is artificial:

- **Skeptical prior:** $P(H_2)$ =0.001
  This value does **not** imply that 1 in 1000 *celestial objects* is artificial — which would be absurd. It means that **among detectable interstellar objects passing through our neighborhood**, we assume that roughly 1 in 1000 could be artificial. This is a deliberately conservative starting point.

- **Moderate prior:** $P(H_2)$ =0.1
  This stance treats the artificial hypothesis as unlikely, but not negligible.

- **Adventurous prior:** $P(H_2)$ =0.5
  A symmetry prior that assigns equal weight to natural and artificial explanations in the absence of strong evidence.

These priors are not claims about the true frequency of artificial interstellar objects; they are **analytical starting points** designed to explore how the evidence shifts belief under different initial assumptions. A full sensitivity analysis later in the appendix examines how the posterior results change when these priors are varied across a wide range.

Likelihood ratios

Let D denote the full evidence set, composed of the seven anomaly clusters and D_i denote each individual cluster anomaly.

For each cluster (D_i), we assign a likelihood ratio comparing an artificial (H_2) hypothesis to a natural one (H_1) as follows:

$$\frac{P(D_i|\,H_2)}{P(D_i|\,H_1)}$$

Assuming conditional independence of these clusters given each hypothesis, the combined likelihood ratio is the product:

$$\Lambda_{total} = \prod_i \frac{P(D_i|\,H_2)}{P(D_i|\,H_1)}$$

The independence assumption is an approximation; in practice, some clusters may share partial causal structure. Because most clusters are neutral (1:1), this approximation has limited impact on the combined likelihood ratios.

The likelihood ratios are not empirical measurements, but structured assessments designed to reflect how well each anomaly cluster fits each hypothesis.

The **likelihood ratio** table below summarizes how easily the evidence fits one hypothesis than the other. A full sensitivity analysis is provided later in the appendix to test how robust the conclusions are to alternative choices of likelihood ratios. The results show that the qualitative outcome remains stable across a wide range of assumptions.

Cluster	Likelihood Ratio (Artificial : Natural)	Rationale
1. Orbital geometry	3 : 1	Retrograde + ecliptic alignment is uncommon but possible
2. Jupiter Hill sphere alignment	10 : 1	Very unlikely by chance; routine for guided objects
3. Composition & chemistry	1 : 1	Unusual but not discriminatory
4. Photometric behavior	1 : 1	Unusual but not discriminatory
5. Jets & non-gravitational forces	10 : 1	Strong, stable jets + smooth acceleration are hard to model naturally
6. Dust morphology	1 : 1	Can be natural or artificial
7. Arrival context	1 : 1	Interesting coincidences but not decisive

Multiplying the cluster likelihoods:

3 * 10 * 1 * 1 * 10 * 1 * 1 = 300

This yields an overall Bayes factor of $\Lambda_{total} \approx 300$

This means the full anomaly stack is **300 times more likely** under an artificial-object model than under a purely natural one — given the conservative assumptions above.

Bayesian updating

We now update three different priors:

- **Skeptical prior:** $P(H_2) - 0.001$
- **Moderate prior:** $P(H_2) = 0.01$
- **Adventurous prior:** $P(H_2) = 0.05$

Bayesian update formula:

$$P(H_2|D) = \frac{P(H_2)\, \Lambda_{total}}{P(H_2)\, \Lambda_{total} + P(H_1)} \quad (1)$$

Where:

- $P(H_2|D)$ is the probability that the object is artificial (H_2), given the full evidence set (D)
- $P(H_2)$ = prior probability of an artificial object
- $P(H_1) = 1 - P(H_2)$ = prior probability of a natural object
- Λ_{total} = combined likelihood ratio (the Bayes factor obtained by multiplying the individual cluster likelihood ratios)

Case 1 — Skeptical prior (0.1%)

$P(H_2) = 0.001$

$P(H_1) = 0.999$

$\Lambda_{total} = 300$

$$P(H_2|D) = \frac{0.001 \times 300}{0.001 \times 300 + 0.999} = \frac{0.3}{1.299}$$

$$P(H_2|D) \approx 0.23$$

Posterior:

- **23% artificial** (H_2)
- **77% natural** (H_1)

Even with a one-in-a-thousand prior, the posterior shifts to nearly one-in-four after seeing the evidence.

Case 2 — Moderate prior (1%)

$P(H_2) = 0.01$

$P(H_1) = 0.99$

$\Lambda_{total} = 300$

$$P(H_2|D) = \frac{0.01 \times 300}{0.01 \times 300 + 0.99} = \frac{3}{3.99}$$

$$P(H_2|D) \approx 0.75$$

Posterior:

- **75% artificial**
- **25% natural**

Under this prior, the artificial explanation becomes the dominant one.

Case 3 — Adventurous prior (5%)

$P(H_2) = 0.05$

$P(H_1) = 0.95$

$\Lambda_{total} = 300$

$$P(H_2|D) = \frac{0.05 \times 300}{0.05 \times 300 + 0.95} = \frac{15}{15.95}$$

$$P(H_2|D) \approx 0.94$$

Posterior:

- **94% artificial**
- **6% natural**

At this point, the natural explanation is struggling to keep up with the evidence.

Bayesian sensitivity analysis

This part provides a compact visual summary of the Bayesian reasoning used in the main text. The goal is not to overwhelm the reader with algebra, but to show — transparently and visually — how the posterior probability responds to different assumptions about the prior and the strength of the evidence.

All three figures apply the same Bayesian update given in Equation 1

To help readers orient themselves, all figures include a **5% posterior threshold**, corresponding to the conventional scientific cutoff for "statistical significance." This is not a decision rule — merely a familiar reference point.

Figure 1 — Posterior vs. Prior (Λ_{total} fixed at 300)

This figure shows how the posterior probability (%) changes as a function of the prior when the likelihood ratio is held constant at 300:1.

Three priors are highlighted:

- $P(H_2) = $ **0.1%** (skeptical)
- $P(H_2) = $ **1%** (moderate)
- $P(H_2) = $ **5%** (adventurous)

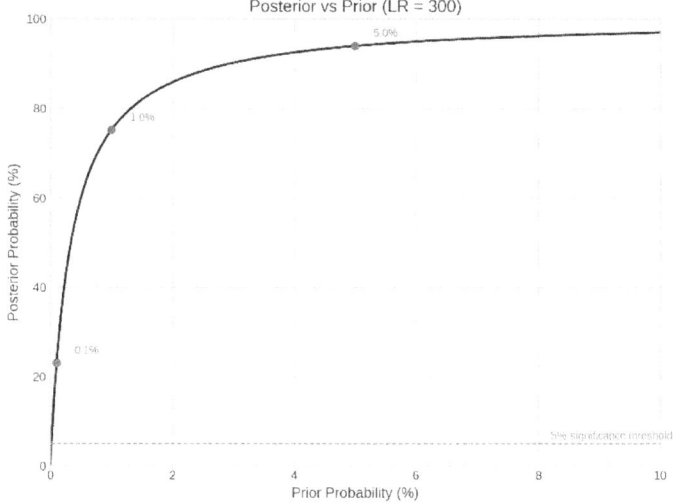

Even the smallest prior crosses the 5% line almost immediately once the evidence is applied. The update is not fragile: a wide range of reasonable priors lead to a posterior that warrants serious consideration.

Figure 2 — Posterior vs. Prior for multiple likelihood ratios

The second figure extends the analysis by showing three curves simultaneously assuming different likelihood ratios:

- $\Lambda_{total} = $ **50** (weak evidence)
- $\Lambda_{total} = $ **300** (conservative evidence)
- $\Lambda_{total} = $ **500** (strong evidence)

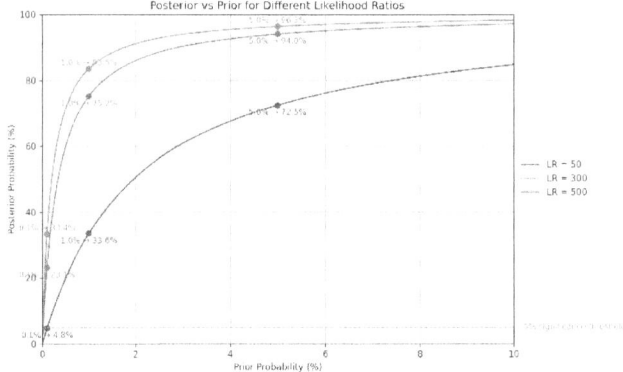

All three rise steeply and cross the 5% threshold for modest priors. The exact Λ_{total} shifts the curve vertically but does not change the qualitative behavior. The posterior is robust across a wide corridor of plausible likelihood ratios.

Figure 3 — Prior vs. Posterior heatmap colored by the required likelihood ratio.

The third figure shows the entire Bayesian landscape at once.

- The x-axis is the prior.
- The y-axis is the posterior.
- The color encodes the likelihood ratio required to move from one to the other.

Only likelihood ratios between 10 and 500 are shown; values outside this range are masked. The heatmap uses an 8-level discrete Cividis palette, chosen for clarity in both color and grayscale.

Green contour lines mark:

Λ_{total} = 10, 20, 50, 100, 300, 500

The dashed line marks the 5% posterior threshold.

This visualization makes two features immediately clear:

- **The significance region is large.**
  A broad swath of the prior–posterior plane corresponds to likelihood ratios well below 300.

- **The update is not sensitive to fine-tuning.**
  Only extremely tiny priors combined with extremely weak evidence keep the posterior below 5%.

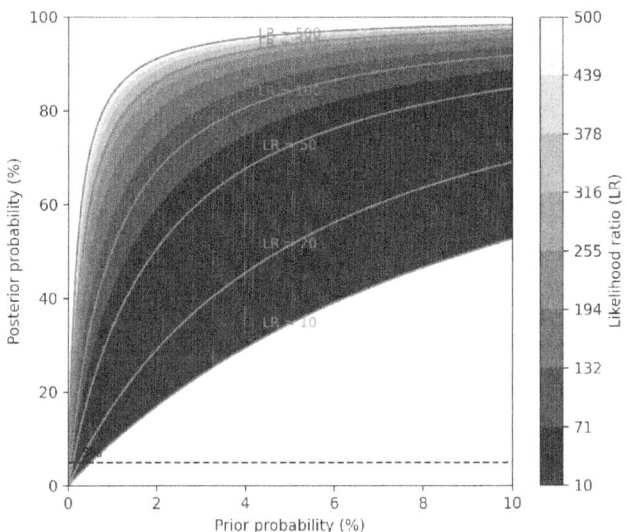

Interpretation

This analysis does not prove that 3I/ATLAS is artificial. What it provides is a transparent, quantitative account of how strongly the full anomaly stack should shift a rational observer's expectations. When:

- the prior is even modestly open to the artificial hypothesis,
- the anomalies are treated conservatively, and
- the evidence is clustered to avoid double counting,

...the combined effect pushes strongly toward the artificial scenario. The natural explanation remains possible — but only by invoking multiple independent "rare but possible" conditions. The artificial explanation requires fewer stretches..

What the Bayesian update shows.

3I/ATLAS exhibits roughly fourteen physical anomalies, but many are related. To avoid inflating the evidence, these are compressed into seven independent clusters. Each cluster receives a conservative likelihood ratio comparing the artificial hypothesis to the natural hypothesis. Most clusters are neutral; only two — the trajectory and the jets — strongly favor the artificial hypothesis.

Under these conservative assumptions, the full anomaly stack fits an artificial-object model about **300 times better** than a purely natural one.

The posterior depends on the prior, but the update is not fragile:

- A skeptical prior of **0.1%** rises to **23%**.
- A moderate prior of **1%** rises to **75%**.
- An adventurous prior of **5%** rises to **94%**.

These results show that, given the observed anomalies, the artificial explanation requires far fewer special assumptions than the natural one.

What the sensitivity analysis shows

The three accompanying figures summarize how the posterior responds to different assumptions about priors and likelihood ratios. All figures apply the same Bayesian update from Equation (1) and include a 5% posterior line as a familiar scientific reference point, not a decision threshold.

Figure 1. The posterior probability is not dominated by any single assumption. A wide range of reasonable priors leads to a posterior that warrants serious consideration.

Figure 2. A wide range of likelihood ratios leads to the same qualitative conclusion. The posterior update remains robust across a broad corridor of plausible evidence-strength assumptions.

Figure 3. The heatmap displays the entire Bayesian landscape. Two features stand out immediately:

- **The significance region is large.** The evidence does not need to be extreme to push the posterior into the "non-negligible" region.
- **The update is not sensitive to fine-tuning.** Only extremely tiny priors combined with extremely weak evidence keep the posterior below 5%. The Bayesian update is robust, not brittle.

Overall conclusion

This appendix is not intended to settle the debate about 3I/ATLAS. Its purpose is to demonstrate that the inference is transparent, mathematically straightforward, and insensitive to reasonable variations in the inputs. The Bayesian framework does not reveal the true nature of the object; it clarifies how strongly the evidence should shift a rational observer's expectations.

Limitations

This Bayesian analysis is intentionally conservative, but it still relies on several simplifying assumptions that should be kept in mind:

- **Independence of clusters.** The seven anomaly clusters are treated as conditionally independent given each hypothesis. In reality, some physical processes may couple multiple observables. Because most clusters carry neutral likelihood ratios (1:1), this approximation has limited impact, but it is not exact.
- **Expert-assigned likelihood ratios.** The likelihood ratios are structured assessments, not empirical measurements. They reflect how well each anomaly cluster fits each hypothesis under current models. Different researchers might assign slightly different values, though the overall conclusion is robust across a wide corridor of plausible ratios.
- **Sensitivity to priors.** Bayesian posteriors necessarily depend on the chosen prior. We explore a range of priors (0.1%–5%) to show how the inference behaves

under different starting assumptions, but readers may reasonably adopt different priors depending on their philosophical stance or background knowledge.

- **Model incompleteness.** Both hypotheses — natural and artificial — are simplified. The natural model may not capture the full diversity of interstellar comet behavior, and the artificial model is agnostic about the intentions, capabilities, or constraints of any hypothetical probe. The analysis compares relative fit, not absolute truth.

- **Post hoc pattern recognition.** Some anomalies were noticed only *after* the object had already drawn attention, which introduces a mild look-elsewhere (you are more likely to notice unusual features simply because you are looking for them) effect. Clustering mitigates this by preventing double counting, but it cannot eliminate all post hoc bias.

- **No claim of proof.** A Bayesian update quantifies how strongly the evidence should shift expectations; it does not establish the true nature of the object. A posterior of 23%, 75%, or 94% reflects degrees of belief under specified assumptions — not definitive classification.

Final Note

This appendix is intended to clarify the logic of the Bayesian update, not to advocate for any particular conclusion about the nature of 3I/ATLAS.

Appendix 2: References and suggested reading

This appendix gathers the most relevant scientific papers, technical reports, books, and online resources related to interstellar objects, techno signatures, and the broader scientific context explored in this book. Entries are grouped by theme for clarity.

Interstellar Object 3I/ATLAS

Breakthrough Listen. *Breakthrough Listen Observations of Interstellar Object 3I/ATLAS.* SETI Institute, 2025.
https://www.seti.org/news/breakthrough-listen-observations-of-interstellar-object-3iatlas/

Sheikh, S. et al. "Radio Observations of 3I/ATLAS Using the Allen Telescope Array." Preprint, 2025.
Referenced in Breakthrough Listen report.

Chandler, C. et al. "Optical Observations of 3I/ATLAS with the Vera C. Rubin Observatory." *Astronomer's Telegram*, 2025.
Referenced in Breakthrough Listen report.

Davenport, J. et al. "Technosignature Search Strategies for Interstellar Objects." 2025.
Referenced in Breakthrough Listen report.

Loeb, A. "Comment on 'Discovery and Preliminary Characterization of a Third Interstellar Object.'" Harvard CfA preprint, 2025.
https://lweb.cfa.harvard.edu/~loeb/atlas_arXiv.pdf

3I/ATLAS Tracker. *Live Interstellar Object Updates.* 2025.
https://i3atlas.com/

1I/'Oumuamua

Meech, K. et al. "A Brief Visit from a Red and Extremely Elongated Interstellar Asteroid." *Nature*, 2017.

Micheli, M. et al. "Nongravitational Acceleration in the Trajectory of 1I/'Oumuamua." *Nature*, 2018.

Bannister, M. et al. "The Natural History of 1I/'Oumuamua." *Nature Astronomy*, 2019.

Seligman, D. & Laughlin, G. "The Feasibility of Detecting Interstellar Objects." *The Astrophysical Journal Letters*, 2020.

2I/Borisov

Jewitt, D. et al. "Initial Characterization of Interstellar Comet 2I/Borisov." *Nature Astronomy*, 2019.

Guzik, P. et al. "2I/Borisov as a Typical Comet from Another Planetary System." *Nature Astronomy*, 2020.

Interstellar Object Population & Dynamics

Do, A. et al. "Interstellar Interlopers: Number Density and Origin of 'Oumuamua-like Objects." *The Astrophysical Journal Letters*, 2018.

Siraj, A. & Loeb, A. "The Mass Budget of Interstellar Objects." *The Astrophysical Journal Letters*, 2019.

Hands, T. et al. "Ejection of Small Bodies from Planetary Systems." *Monthly Notices of the Royal Astronomical Society*, 2019.

Technosignatures & SETI Context

Freitas, R. & Valdes, F. "A Search for Natural or Artificial Objects Near the Earth." *Icarus*, 1985.

Sheikh, S. et al. "A Framework for Technosignature Searches." *Acta Astronautica*, 2020.

Wright, J. "The Search for Extraterrestrial Technosignatures." *Annual Review of Astronomy and Astrophysics*, 2021.

General Audience Books

Avi Loeb — *Extraterrestrial: The First Sign of Intelligent Life Beyond Earth* (2021)

Caleb Scharf — *The Zoomable Universe* (2017)

Carl Sagan — *The Demon-Haunted World* (1995)

David Grinspoon — *Earth in Human Hands* (2016)

Alan Stern & David Grinspoon — *Chasing New Horizons* (2018)

Scientific & Technical Background

Meech, K. et al. "Interstellar Objects: A New Frontier." *Annual Review of Astronomy and Astrophysics*, 2020.

Trilling, D. et al. "Implications of Interstellar Objects for Planetary System Formation." *The Astronomical Journal*, 2018.

Siraj, A. & Loeb, A. "Interstellar Objects as Probes of Exoplanetary Systems." *The Astrophysical Journal Letters*, 2020.

Philosophy of Science & Uncertainty

Thomas Kuhn — *The Structure of Scientific Revolutions* (1962)

Nassim Nicholas Taleb — *The Black Swan* (2007)

Ian Hacking — *The Taming of Chance* (1990)

Probability, Statistics & Bayesian Reasoning

- Spiegelhalter, D. The Art of Statistics: How to Learn from Data. Basic Books, 2019.

- Hartshorn, S. Tell Me the Odds: A 15 Page Introduction to Bayes Theorem. Fairly Nerdy Publishing, 2017.

- Kurt, W. Bayesian Statistics the Fun Way: Understanding Statistics and Probability with Star Wars, LEGO, and Rubber Ducks. No Starch Press, 2019.

- Ellenberg, J. How Not to Be Wrong: The Power of Mathematical Thinking. Penguin, 2014.

Online Resources

Breakthrough Listen (SETI Institute)
https://www.seti.org

3I/ATLAS Live Tracker
https://i3atlas.com

Harvard CfA Interstellar Research
https://www.cfa.harvard.edu/

Sheikh, S. et al. "Radio Observations of 3I/ATLAS Using the Allen Telescope Array." Preprint, 2025.
Referenced in Breakthrough Listen report.

Chandler, C. et al. "Optical Observations of 3I/ATLAS with the Vera C. Rubin Observatory." *Astronomer's Telegram*, 2025.
Referenced in Breakthrough Listen report.

Davenport, J. et al. "Technosignature Search Strategies for Interstellar Objects." 2025.
Referenced in Breakthrough Listen report.

Loeb, A. "Comment on 'Discovery and Preliminary Characterization of a Third Interstellar Object.'" Harvard CfA preprint, 2025.

Acknowledgments

This book would not exist without the curiosity, conversations, and encouragement of many people. I want to thank the broader scientific community whose open data, published analyses, and ongoing debates made it possible to examine 3I/ATLAS and the other interstellar visitors with rigor and transparency. Science advances through collective effort, and this work stands on the shoulders of that shared endeavor.

I would particularly like to recognize the communicators Neil deGrasse Tyson, Avi Loeb, Derek Muller (Veritasium), and Javier Santaolalla (Date un Voltio) for their commitment to keeping scientific communication honest, open, and accessible. Although I don't know them personally, their work was an inspiration for the development of this book.

In writing this work, I made extensive use of AI-assisted drafting and organizational tools. They helped me explore alternative phrasings, test explanations, and maintain coherence across the manuscript. These tools were valuable aids, but the interpretations, reasoning, and conclusions presented here are entirely my own.

To my family, whose patience and support sustained me through long nights of writing and revision — thank you. Your belief in me has always been the quiet force behind every project I undertake.

Finally, I am grateful to every reader who approaches this topic with an open mind. Curiosity is humanity's oldest tool, and it remains our most powerful one.

About the Author

Dr. E.A. Gálvez is a scientist and science author whose work combines rigorous analysis with accessible communication. Trained in the physical sciences and experienced in interpreting complex natural systems, he brings a multidisciplinary perspective to questions at the intersection of astronomy, physics, and planetary science.

His research background spans geophysical modeling, data analysis, and the study of natural anomalies—experience that informs his approach to interstellar visitors and the scientific puzzles they present. He is committed to clear, evidence-based reasoning and to making technical subjects understandable to a broad audience.

Born in Mexico and educated in both Mexico and the United Kingdom, he holds a Ph.D. in geophysics and writes in both English and Spanish. E.A. Gálvez is dedicated to expanding access to scientific ideas across cultures and languages. *The Interstellar Visitors* is his first book.

Printed in Dunstable, United Kingdom

79660034R00137